From Geocentric to Heliocentric: How Discoveries Are Made

"This book allows us to rediscover the inspiration from nature and to trace the meditation process of the ancient philosophers, from the perspective of our ancestors. Mr. Ge is neither a professional astronomer nor a historian, which enables him to start from the perspective of an ordinary person. This is an insightful book, which can be easily understood by a high school student, since not much background knowledge is required. The book tells us not only the result, but also the exploration process, which stimulates our interest in nature and enables us to meditate, the ability more important than knowledge in the exploration of the universe."

—Researcher Chen Xuelei, *Chief scientist in the research group of dark matter and dark energy at the National Observatory*

"In the history of science, many prophecies that were made through careful analysis, rigorous inference and fertile imagination emerged far ahead of the actual observations, so did human discovery of the universe. This book deserves a careful look."

—Professor E Weinan, *academician from Chinese Academy of Sciences, professor from Peking University*

"This book tells a profound "story" in plain language—how human beings gradually understand nature through observation and mathematical construction. Modern science originates from the knowledge of heaven and earth. Readers can learn from this book the spirit of science and the power of positivism."

—Professor Qian Hong, *Department of applied mathematics, University of Washington, Seattle, USA*

"The vast majority of popular science books tell us, "what the world is" and "why", unlike this book which explains the exploration process in plain language. This is a meaningful book because the exploration process reflects the meaning and spirit of science, as well as the wisdom of scientists, which makes this book especially suitable for high school students and interested common readers."

—Professor Li Weigu, *School of Mathematical Sciences, Peking University*

"In addition to the rich scientific knowledge, the author, by tracing the history, tells a great truth in philosophy of Science: science develops by "falsification" rather than "verification"! "Verification" is just a factor in development, but "falsification" is the motivation for development."

—Professor Zhang Ping, *Department of East Asian studies, Tel Aviv University, Israel; Yangtze Fund Scholar and Chair Professor in China*

Yunbao Ge (葛云保), born in Danyang City, Jiangsu Province, China in 1949. He graduated from Hohai University in 1976. He is a civil engineer and astronomy enthusiasts.

Yunbao Ge (葛云保)

From Geocentric to Heliocentric: How Discoveries Are Made

科学出版社
Science Press

Springer

Yunbao Ge (葛云保)
Beijing, China

Translated by
Zhaocan Liu (刘兆灿)
School of Foreign Languages
Gannan Normal University
Ganzhou, China

Qian Xu (徐倩)
Law School
Shanghai International Studies University
Shanghai, China

Hao Ge (葛颢)
Beijing International Center for
Mathematical Research and Biomedical
Pioneering Innovation Center
Peking University
Beijing, China

ISBN 978-981-16-7778-6 ISBN 978-981-16-7779-3 (eBook)
https://doi.org/10.1007/978-981-16-7779-3

Jointly published with Science Press, Beijing, China
The print edition is not for sale in China (Mainland). Customers from China (Mainland) please order the print book from: Science Press.

This Springer imprint is published by the registered company Springer Nature Singapore Pte Ltd.
The registered company address is: 152 Beach Road, #21-01/04 Gateway East, Singapore 189721, Singapore

Preface

Today, heliocentrism is widely known, and it seems to be a self-evident truth. But is this really the case? For thousands of years, what human beings have seen with their own eyes is that the sun, the moon, and stars move around the earth.

Who has seen the earth revolve around the sun?

Nobody.

So do you still believe the earth really revolves around the sun?

Mr. Ge Yunbao asks such questions in this book *From Geocentric to Heliocentric: How Discoveries Are Made*.

This is a unique popular science work. Its purpose is not to introduce the knowledge of heliocentrism, but to guide us to think and review the scientific process of establishing heliocentrism. To make an inappropriate analogy, this book is like a time machine, carrying readers across the time to rediscover the laws of heaven and earth. It takes us back to the ancient times of mankind. Our ancestors saw changes of seasons and the sun's rising and setting, and finally discovered the relationship between the sun and four seasons. It takes us back to ancient Greece, watching the celestial phenomena at night, measuring the sun's shadow at day, finding that the earth is spherical, and calculating the size of the earth. It takes us to follow talented Aristarchus to measure the size of heaven and earth, and push the earth down the altar in the center of the universe, which surpasses the age for two thousand years. It takes us to be with familiar or unfamiliar sages such as Apollonius, Hipparchus, and Ptolemy, and constructs the grand system of geocentric theory with beautiful circular motion and exquisite geometric model, which prevailed for thousands of years. It takes us to the 16th century, together with Copernicus, to clear away the dark clouds in Ptolemy's sky, and to find that the universe in heliocentrism is so simple and harmonious. It takes us to break the old law of circular motion with the "Sky Legislator", discover new satellites and continents together with "Columbus of Sky", and interpret the unified world together with Newton's mechanics and universal gravity. It leads us to observe from the perspective of ancient people, appreciate the hardships of scientific exploration and enjoy the joy of scientific discovery in thousands of years.

This is a unique book. Its author is an astronomy amateur instead of an astronomer. Mr. Ge spent nearly ten years studying, observing, calculating, and thinking, and finished this book, just to answer one question: Why do humans believe that the earth revolves around the sun? For this reason, this is a popular science book for the public. To understand it, you just need simple mathematics and reasoning. For this reason, this book is full of emotions, and you will feel the author's interest in science and passion for exploring science in the words. For this reason, it is a book that inspires people to think and explore, which is the most recommended in my opinion. As my good friend, Professor Chen Xuelei of National Astronomical Observatory, said in his recommendation, "The interest in the mysteries of nature and the ability to think are more important for exploring the universe than the latest concrete knowledge".

That is right, scientific knowledge will make progress in pace with the development of mankind. Just two years ago (March 2013), the Planck cosmology probe first confirmed that the sun moves at a high speed of about 380km/s in the embers of the Big Bang. The title of this paper is somewhat strange: "Eppur si muove", which means that "It still moves". It is said that this was Galileo's mumbling to himself after he was forced to give up heliocentrism in 1633. The "absolute" movement of the whole solar system in astrospace has gone far beyond the framework of heliocentrism. But it is still titled "Eppur si muove", which expresses the respect of contemporary astronomers to the sages of science, and it is a cheer for them to clear the fog and reveal the mysteries of heaven and earth.

What is eternal is that the spirit of scientific exploration and the light of scientific wisdom can shine everywhere. Scientists think so, science enthusiasts think so, and this book thinks so.

April 2015

Zhang Pengjie (张鹏杰)
Department of Physics and Astronomy
Shanghai Jiao Tong University
Shanghai, China

Acknowledgements

It has long been a wish of mine to translate my book into English and publish it abroad, so that it can be read by more university and high school students, as well as those who love science. Today, that wish has come true. I am very happy!

I would like to express my gratitude to all those who helped me in the translation and publishing of this book.

First, my deepest gratitude goes to Ludwig Arnold, a famous German mathematician, for introducing and recommending my book.

Next, I would like to thank Liu Zhaocan (刘兆灿) from Beijing Huilan International Culture Communication Co., Ltd., who took painstaking effort to translate my book into English.

I would also like to thank Dr. Xu Qian (徐倩) from Shanghai International Studies University for reviewing and revising the English manuscript over and over again.

Then, my heartfelt thanks go to Asso. Prof. Ge Hao (葛颢) from Peking University, for his great efforts to translate and publish this book, and for reviewing and revising the translation over and over again.

Finally, I would like to express my gratitude to Science Publishers and the editors of Springer, for their hard work in bringing this book to fruition.

I could not imagine the publication of this book without their help. Many thanks to them!

Introduction

Who has ever seen the earth revolve around the sun?

No one has ever done it. We know from many astronomical books that the earth's orbit around the sun is about 300 million kilometers in diameter. Imagine, if we want to see the revolution of the earth around the sun with our own eyes, we must stand in one place perpendicular to the earth's orbital plane, hundreds of millions of kilometers away from the center of the sun, and watch it for a whole year. It's just like looking at a huge clock, we must stand in front of the clock and be far enough from it to have a panoramic view of the huge clock. So far, human beings are far from being able to do this.

So do you still believe that the earth really revolves around the sun? Or have you ever wondered that since no one has seen the earth revolve around the sun, who said that the earth revolves around the sun? How did he know it? What reason did he have to say so? Why do other people believe it?

Will you be surprised if I tell you, as early as four or five hundred years ago, when human beings could not leave the earth at all, Copernicus had demonstrated that the earth is a planet in the solar system, and it is constantly revolving around the sun besides its rotation, despite the fact that science and technology were still underdeveloped at that time, and even telescopes were not invented? Will you be more surprised if I continue to tell you, as far back as more than 2000 years ago in ancient Greece, when the range of human's activities was very limited, some people could point out that the earth was spherical without any instrument, and some people later argued that the earth should be revolving around the sun rather than the sun revolving around the earth?

Under China's current educational model and traditional educational philosophy, teachers usually only tell students what the world is like. In other words, what teachers tell students is the result of predecessors' exploration. But how our predecessors explored these results and what they thought at that time are rarely mentioned. Most of the popular science books we see in libraries and bookstores, including popular science books on astronomy, tell readers the achievements of previous exploration of predecessors one by one. And they at most tell you some simple history, such as something discovered by someone, some theory created by someone, and so on. But they rarely mention what the great scientists thought at that time and how they found these answers, and it is not systematic even if they do. Therefore, although we know from the books in class that the earth is spherical, the sun is the center of the solar system, the earth is a planet of the sun, which revolves around the sun like several other planets, the solar system is one of tens of thousands of star systems in the Milky Way, and there are countless Milky Way in the universe... But how do astronomers know all of these? I believe many people know little about it.

It is surely important to understand and master the results of predecessors' exploration, but is that enough? I think we should also properly understand the exploration process of our predecessors, which is also very important, because in this process, we can see how those talents thought, analyzed, guessed, judged, reasoned, questioned, and proved, how they found problems that ordinary people cannot find, and how they solved problems that ordinary people cannot solve. Much of the scientific common sense today contains the great thoughts and unparalleled wisdom of scientific talents! This can undoubtedly enlighten and help us to explore the new unknown world.

Let's make an analogy. This book is similar to explaining a geometric proof question in junior high school. What are the known conditions? It is the vast sky before our eyes, which is open to all people. What is the result to prove? It is what the universe really looks like. Although this result has long been well known, many people do not know this interesting process of exploration.

Here I would like to explain to the readers that the universe mentioned in this book refers to the solar system. In fact, in the age of low science and technology, what astronomers could see were unchanging celestial sphere and the sun, the moon, and stars that move slowly in the background of the celestial sphere. Therefore, for quite a long time, human beings thought that the solar system was the whole universe. From the perspective of understanding the laws of things, human beings must first know the solar system, and then gradually the whole universe.

Human beings knew very little in ancient times, and they have come a long way from this to gradually realize that the earth is spherical, and then the earth is one of the planets of the sun. During this long journey, human beings only relied on their own naked eyes and some simple instruments. Besides, the knowledge they need is just the geometry knowledge of junior high school, mainly the knowledge of parallel lines, circles, and triangles. Therefore, this history should be easy to understand for a person with junior high school knowledge and spatial imagination. This book is written for those who have a certain knowledge of geometry and spatial imagination, and have learned a lot of astronomical knowledge in class and books, but have never carefully observed the sun, moon, and stars, and only occasionally have curiosity and doubts about the sky.

Let's review this long, tortuous, and exciting journey of verification together.

Contents

1 The Milky Way with Bright Stars 1
1 The Sky Is Like a Dome Covering the Wild Land 1
2 The Sunrise and Sunset 2
3 Twinkling Stars 10
4 A Bright Moon 13
5 Unquiet Sky 18
6 Eyes and Mind Combined 20

2 The Motivation for Observing the Astronomical Phenomena at Night 23
1 The Urgent Needs of Daily Production and Life 23
2 Comprehending God's Will 35
3 Just Out of Curiosity 40

3 Mysterious Whereabouts of the Sun, Moon and Stars 43
1 Where Do the Stars Hide? 43
2 Where on Earth Is the Sun? 49
3 Variations of the Moon 53
4 How Far Are They from Us? 56

4 Puzzling Earth 59
1 A Big Obstacle 59
2 The Earth Is Under Our Feet 62
3 How Big Is This Sphere? 68

5 Guess What the Universe Is Like 71
1 Who Can Guess What the Universe Is Like? 71
2 A Fire in the Center of the Universe 73
3 The Earth in the Center of the Universe 76
4 The Sun in the Center of the Universe 83

6 Geocentric System Prevailing for a Thousand Years 89
1 Magical Epicycle-Deferent System 89
2 Let the Earth Deviate Slightly from the Center 96
3 A Complete Geocentric Model 102
4 Questioning Ptolemy's Geocentric System 110

7 Put the Sun in the Center of the Universe 113
1 New System Arrived on the Scene 113
2 Voice of Doubt 119
3 Totally Enlightened by One Move 121

8 Combination of the Two Major Systems 135
1 Eclectic Tychonic System 135
2 Bridge Between Two Major Systems 137
3 Can History Be Assumed? 143

9 Towards Victory 147
1 Sky Legislator—Kepler 148
2 "Columbus" of Sky—Galileo 155
3 Newton Standing on the Shoulders of Giants 161
4 Evidence of the Earth's Revolution 164

Epilogue 169

Bibliography 173

Chapter 1
The Milky Way with Bright Stars

1 The Sky Is Like a Dome Covering the Wild Land

Human beings must have been awed by the vast and mysterious stars when they looked up at the sky for the first time, and since then they have begun their endless research and exploration.

Astronomy is the science of observation. Humans had observed and studied celestial bodies with naked eyes for a rather long time before the telescope was invented in 1608.

Living in the concrete forest far from nature, people in the metropolis, including those high school students who should be most curious about nature, rarely have the opportunity, ability or mood to enjoy the sky. Many of them have seldom seen the sunrise or sunset, and they also pay little attention to the moon, let alone the rises and falls of it. They cannot see stars because the starry sky has been dimmed by the bright lights in the city and the polluted atmosphere. Sometimes, a small number of high school students are capable of witnessing the sunrise and sunset, changes of the moon and the stars when they make a trip to the seaside, grassland and countryside. But it's just a glimpse without attention, curiosity, questions or thoughts.

Only children living in rural areas are likely to observe the sky, but most of them just occasionally have a look at it, knowing little about astronomical phenomena, due to their study stress and the fast pace of modern life.

Therefore, it is necessary for us to make up the knowledge of astronomy and tell people what we can see with our naked eye in detail. We cannot understand how people thought about astronomical phenomena in the past if we know nothing about what they have observed. We have to get a clear understanding of the known conditions first, before comprehending how such a geometric problem is proved.

What can we see in the sky with only our naked eyes? We need to make the following hypotheses to illustrate this point. First, suppose we are observing the sky in mid-latitudes of the northern hemisphere, because most ancient human civilizations were in this area. Second, suppose we are observing the sky from a place that is wide and flat and far away from the city, because it's impossible for us to make

Y. Ge, *From Geocentric to Heliocentric: How Discoveries Are Made*,
https://doi.org/10.1007/978-981-16-7779-3_1

an observation of the sky in the ravine or modern city where there are too many barriers and lights and the horizon is almost invisible. When we are far away from the modern illuminated city, we'll have a better sense of how the sky influenced ancient civilization. Third, suppose the sky is always clean without pollution as what we are studying is not the weather but the celestial bodies which we cannot see on rainy or smoggy days. Fourth, suppose we are having a consecutive observation for a long time, such as one day, one month or even over one hundred years, because the apparent motion of celestial bodies is so slow that a long-term observation is essential if we want to find out the laws of their motions and changes. Fifth, suppose we are observing the sky from two distant places at the same time when necessary, because many astronomical phenomena have to be compared through observing from different locations. Last, suppose we make the observation of sky through the ages—sometimes we do it with astronomers of this era, sometimes with astronomers from a certain period in ancient times, and sometimes with ordinary people in remote antiquity.

Now, let's observe the sky.

We can see that the sky and ground are combined into a flat straight line which is called skyline if we stand in a flat and open place and look at a distance. When we slowly turn our eyes to the left and again, and again, then it's not hard to find that the horizon is a circle around us in fact, and it is certainly a giant one which is called the horizontal circle by astronomers. As is described in a Chinese ancient poem, "The sky is like a dome covering the wild land", in another word, the sky is like an arched tent high in the middle and drooping around, or like half a huge spherical shell, covering the vast land. Besides, you'll feel yourself at the center of the huge hemisphere wherever you stand. (Fig. 1).

2 The Sunrise and Sunset

There is no doubt that ancient people, no matter which continent or population they belonged to, must have started to know the celestial bodies with the sun as its beginning. As the sun is so closely related to human life, we'll also begin with the sun.

How big is the sun? The ancient people had no idea how far away the sun was or how big it was, nor did they know how far or how big other celestial bodies were, let alone the distance between one another. Therefore, they could only express the size and distance of celestial bodies by the angle. Around 2000 BC, the ancient Babylonians divided the circumference into 360°, 1 degree into 60 arc min, 1 arc min into 60 arc sec. The method has been in use till now for more than 4000 years. The angular diameter of the sun is about half a degree when measured by this standard. It would take about 720 suns to form a circle around us on the horizon one by one. Scientists usually use the angle to express how high the sun is or how far the moon is from the sun, and this is what we will often mention later, i.e. the "angular distance" or "apparent distance".

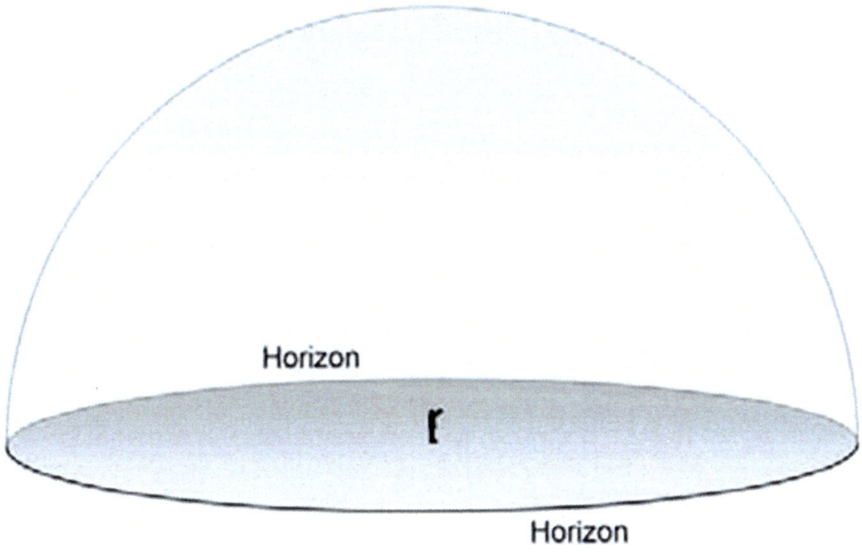

Fig. 1 The sky is like a dome covering the wild land

The shining sun brings us light and warmth, and it has a regular motion: it rises from the east, and it moves higher and higher over our heads before it gradually sinks to the west. The sky becomes darker and dimmer slowly as the sun falls below the horizon. After the night, the darkness will gradually disappear, then the sun will rise from the east and pass over our head again, and set to the west at length. Round and round it goes and never stops. People still believe the sun is rising when the sky starts to brighten even if it is a rainy morning. In ancient times, nothing could prevent people from seeing the rising and falling of the sun except for mountains and clouds.

In the beginning, the interval between the two successive rises of the sun was called "one day", and today we regard midnight as the dividing point of one day. This must be the earliest unit of time that human beings knew, and it is the most basic unit of time. The longer unit of time known later must be measured by the "day".

We will find that the sun does not rise at the same point on the eastern horizon every day as long as we observe it for some time. Suppose we are facing the direction where the sun rises (Fig. 2) on a spring morning, and see the sun rising just behind a large tree (the sun behind a large tree in Fig. 2). The next day, the sun seems to rise behind the tree, but the rising point of the sun has moved a little to the north, about a sun's diameter which equals half a degree (The sun on the left of the large tree in Fig. 2). The movement of the sun becomes apparent after seven or eight days because it has risen from the north of the tree. The rising point moves north slowly just like this, farther and farther away from the tree, but there is a limit to this movement (the sun at the left edge in Fig. 2), at which the rising point will slowly move back towards the tree, over it, and towards the south. Similarly, there is also a limit in the south

Fig. 2 Displacement of sunrise in two adjacent days in spring and two limiting positions

(the sun at the right edge in Fig. 2), where the rising point will move back slowly again.

What's the distance between the south and north that the rising point covers? It's forty to fifty degrees if seen in the middle latitudes of the northern hemisphere. So, when we say the sun rises from the east, the "east" is a general direction. The midpoint between the two points of south and north is the due east. In a complete cycle of the north-south round trip, the sun rises from the due east for only two days: the "Vernal Equinox" and the "Autumn Equinox".

The setting point of the sun, just like its rising point, has imperceptible variation every day. It also repeats its slow forward and backward movement from south to north and from north to south within a certain range.

Rural children should know this much better. Urban children who live in a high building with a window facing the east or the west can find this movement as long as they observe it for ten days or half a month. Of course, it's better to do it for a longer time. The two sunset photos in Fig. 3 were taken at the same place but on two evenings in winter and summer respectively. We can see that the positions of the sunset are very different.

The sun moves across the sky in a large circular arc from rising to setting. People in the middle latitudes of the northern hemisphere can find that the arc of the sun's trajectory slopes southward (Fig. 4), by watching the movement of the shadows of trees or buildings and by observing the light through the doors and windows facing south. The ancient people perhaps found this in the sunlight through the cave facing south.

Imagine there is a straight tree growing on the flat ground. It casts a long shadow on its west when the sun rises in the morning. As the sun goes up higher and higher, the tree's shadow gradually shortens and its direction also changes slowly. The shadow turns shortest when the sun is at its highest point of the day, pointing to the north. After this moment, the sun moves to the west lower and lower with the tree shadow getting longer and moving to the east gradually.

Fig. 3 Sunrise in winter and summer

Does the shadow of the tree change like this every day? Yes, but not exactly so. As we have said before, because the rising position of the sun changes day by day, the starting direction of the shadow of the tree changes along with it, too. For example, if we can set up a tall upright pole in the middle of the flat ground, it will cast a shadow on the ground after sunrise. If we draw the shadow of the pole on the ground every ten minutes, at the end of the day, there will be a beautiful pattern formed by many straight lines on the ground.

Figure 5 shows three patterns of shadows of a pole drawn on three specific dates.

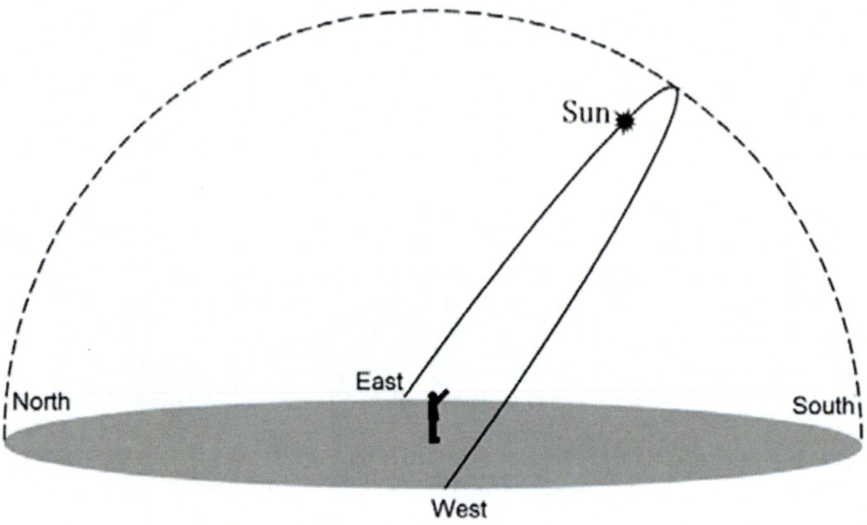

Fig. 4 The arc formed by the sun's trajectory inclining to the south

It is a pattern seen from above, also known as top view. Observation site: 115 °E, 35 °N. The solid line in Fig. 5 is the shadow of the pole at several time points during the day, and the dotted line is the boundary within which the shadow of the pole changes.

We can find how the shadow of the pole changes by comparing these three patterns. In a day, not only does the shadow always change from long to short and then to long, but also its direction changes from west to east gradually, and the pattern formed by the shadow is perfectly symmetric, whether it is on Winter Solstice or Summer Solstice, Vernal Equinox or Autumn Equinox. These are the commonalities of all three images and of changes of shadow on any day of the year. And we will find the fourth commonality if we observe it more carefully: On any given day, the shadow of the pole always points precisely to the north when it is the shortest, which is at midday. This is a significant characteristic. However, the overall shapes of these three images vary a lot. For instance, the shadow originally points west by north on the Winter Solstice and originally points west by south on the Summer Solstice. Besides, the length of the shadow varies on different days. Figure 6 shows a pole perpendicular to the ground at 35 °N, 115 °E, and there is a considerable difference in the length of its shadow at noon on the Winter Solstice and the Summer Solstice.

These are the changing characteristics of the height and position of the sun. It is easy for people to find that the height and position of the sun change gradually from morning to night. Comparing the same time of every day throughout the year (take 10. am as an instance), you will discover that the height and position of the sun also change from day to day, but the sun only changes in height at noon every day and is always in the due south of the sky.

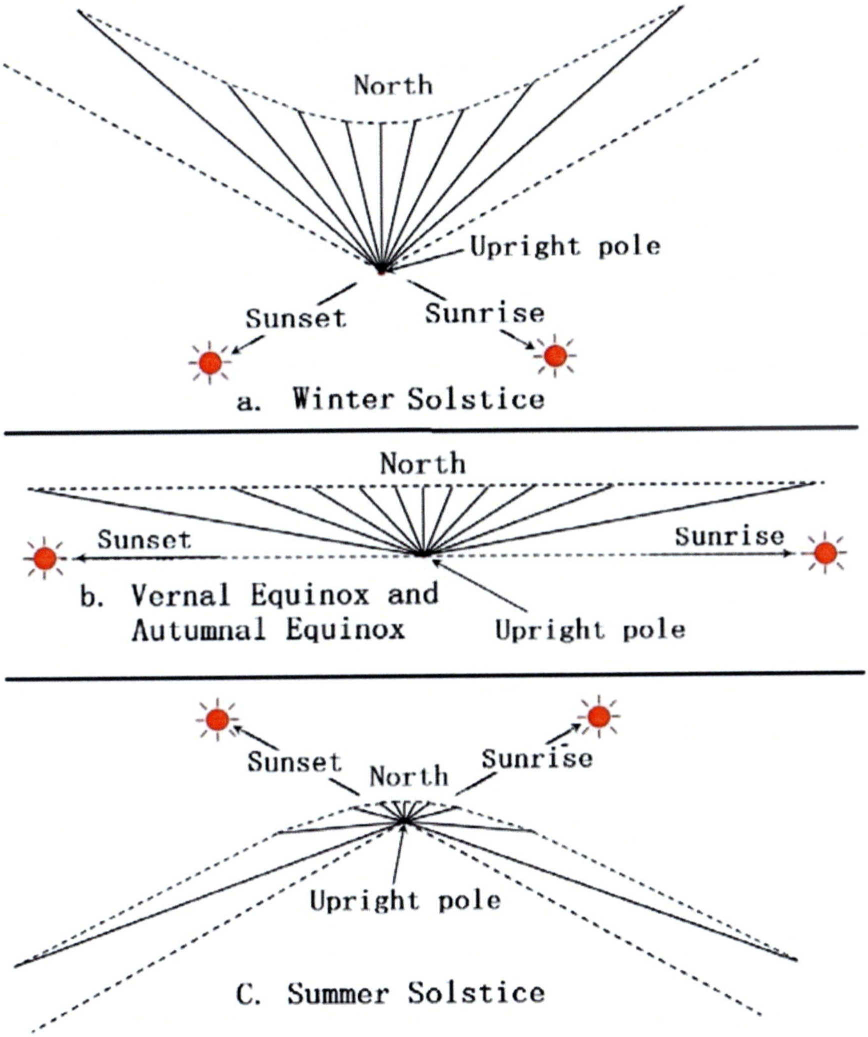

Fig. 5 The pattern scanned by the shadow of the upright pole

The sun rises earlier and earlier and sets later and later as the position of its everyday rising and setting moves from south to north, that is to say, the daytime gradually becomes longer and the nighttime shorter. On the contrary, the sun rises later and later and sets earlier and earlier as the position of its everyday rising and setting moves from north to south, that is to say, the daytime gradually becomes shorter and the nighttime longer. Many of us have such an experience. In summer, it had already dawned when we went to school in the morning, and it was not dark yet when we came home from school, so we could play outside for a long time. However, in winter, it was still dark when we went to school in the morning, and it was already

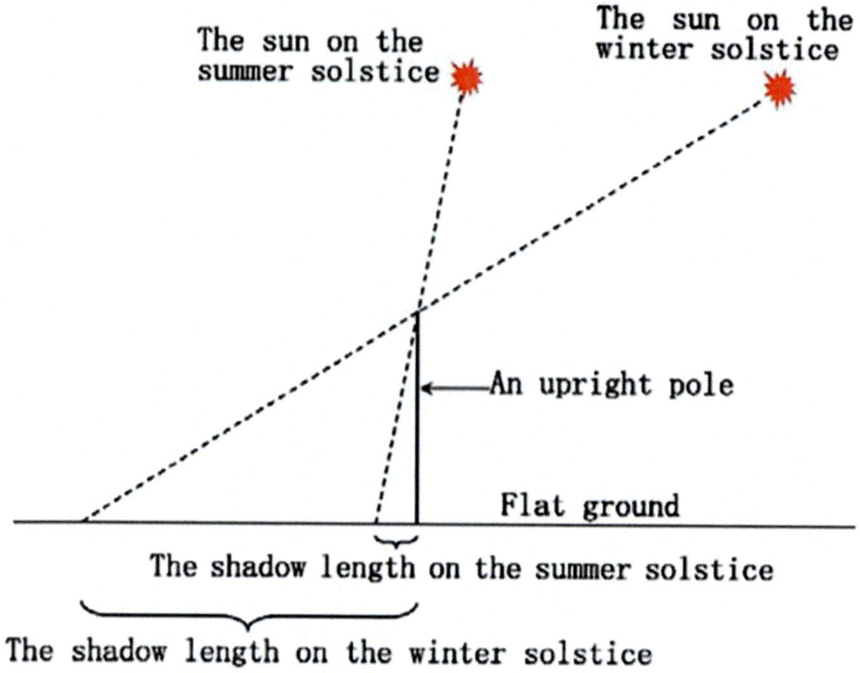

Fig. 6 Shadows of the pole on winter Solstice and summer Solstice

dark when we returned home. In a complete cycle where the sun moves to the south and back to the north, there are only two days when day and night are equally long: the "Vernal Equinox" and the "Autumn Equinox".

The points of sunrise and sunset are fully in step with the height of the midday sun (which is equivalent to the shadow of the pole at noon) and the length of day and night: The sun will set at the northernmost point of the west when it rises from the northernmost point of the east, and it will draw a northernmost arc in the sky. At noon, the sun reaches its highest point in due south and is the closest to us, when the length of the shadow of the pole is the shortest. The date of the longest day and the shortest night is called "Summer Solstice" because it is during the hot summer. By the same token, the sun will set at the southernmost point of the west when it rises from the southernmost point of the east, and it will form a southernmost arc in the sky. At noon, the sun reaches its lowest point in due south and is the farthest over us, when the length of the shadow of the pole is the longest. This date of the shortest day and longest night is called "Winter Solstice" because it falls in the coldest winter. Figure 7 shows the track of the sun rising from the east and setting in the west on the four specific days, namely, Summer Solstice, Vernal Equinox, Autumn Solstice and Winter Solstice.

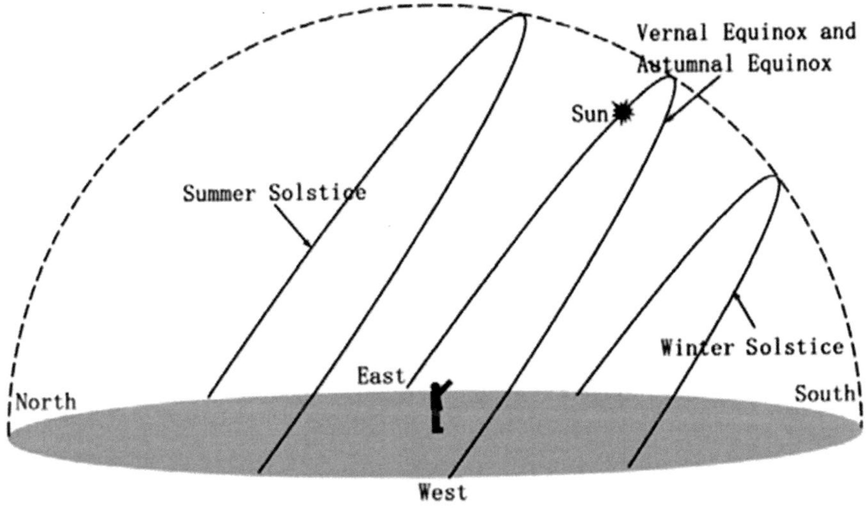

Fig. 7 The trajectories of the sun rising in the east and setting in the west on four specific days

The changes of apparent motion of the sun make a major cycle, from the winter solstice to the next winter solstice, or from the summer solstice to the next summer solstice.

We'll see that the farther north we reach, the more slanting the arc formed by the sunrise in the east and sunset in the west is. So, we can notice that the sun in Nanjing looks closer over us than that in Beijing at noon on the same day.

As has been mentioned before, the length of day and night is not invariably equal though there are twenty-four hours in each day. The length of day and night are dissimilar in a year except for the two equinoxes. The days are shorter and the nights are longer in winter while the days are longer and the nights are shorter in summer. If we continue to move the observation point, we can further discover that on the same day, the farther south we go, the less difference there is in the length of day and night, and the farther north we go, the more difference there is in the length of day and night. For example, on New Year's Day in Beijing, the sun rises at 7:36 and sets at 16:59, and the daytime is 9 h and 23 min and the night is 14 h and 37 min, so there is a difference of over five hours between day and night. On New Year's Day in Nanjing, the sun rises at 7:06 and sets at 17:11, and the daytime is 10 h and 5 min and the night is 13 h and 55 min, so the difference between the day and night is less than four hours. The winter nights in the north are therefore very long. Likewise, on July 1st, the sun rises at 4:50 and sets at 19:47 in Beijing, and the daytime is 14 h and 57 min and the night is 9 h and 3 min, so there is nearly 6 h' difference between the day and night; while in Nanjing, the sun rises at 5:03 and sets at 19:15, and the daytime is 14 h and 12 min and the night is 9 h and 48 min, so the difference between the day and night is still less than 4 h.

These are the movements of the sun that we can see directly with our naked eyes. Of course, besides the short period of sunrise and sunset, we can directly watch the sun on foggy days or when the sky is covered by thin clouds. But we are not able to do it during the rest of the day unless we take some proper measures. We can judge the location and height of the sun by the shadow of trees in the sun or the site where the sunlight shines into a cave.

3 Twinkling Stars

As daytime animals, the ancient people worked from sunrise to sunset and the night sky was less related to their daily life. The sky at night always seems the same for most of us, with countless stars scattering in it. Today, in a city that is becoming ever bigger, taller and brighter, people can hardly see the stars even if they want to.

The starry sky at night, where bright and pale stars scatter all over at random, is gorgeous. It is like innumerable flickering eyes that are staring at us.

Let's look up around us from somewhere far away from city with a wide view at 21:00 on a clear night. There are incalculable stars in the semicircular sky, and we will find its secret on condition that we carefully observe the sky for a longer time: Facing the east, we'll see that stars are slowly rising one by one and higher and higher; looking up at the sky, we will see that stars are slowly moving in turn from east to west; turning to the west, we will see that stars are slowly falling and sinking below the horizon one after another. It turns out that stars rise and fall like the sun and moon! This may not have occurred to a good many people who never pay attention to the night sky.

The whole sky is moving from east to west and nearly half the stars will sink below the western horizon in six or seven hours, while the stars rising from the east will fill about half of the sky.

All the stars can be seen revolving counterclockwise around some point slowly when we face due north. Figure 8 presents photos that came into being after continuous exposure to the northern sky with a fixed camera, by which we can feel the rotating sky. However, as an observer, the revolution of stars we see every night is not a complete circle, but just nearly half of it.

We can also pay special attention to the well-known Northern Dipper. Figure 9 shows the Northern Dipper observed at different times on the same night, in which we can find that it is revolving slowly counterclockwise around some point called the North Pole. And the North Star is close to the North Pole and easy to locate, although it is not bright enough. It seems that only the North Star stays in that position forever in the whole sky.

The stars close to the North Pole make a smaller circle when they revolve while those far from the North Pole make a larger circle. But they revolve at exactly the same angle during the same period of time, whether they are far from or near the North Pole. Seen from the middle latitudes of the northern hemisphere, an area around the North Pole can be seen lying above the horizon at any time of night while stars

Fig. 8 The northern starry sky revolving around the north pole

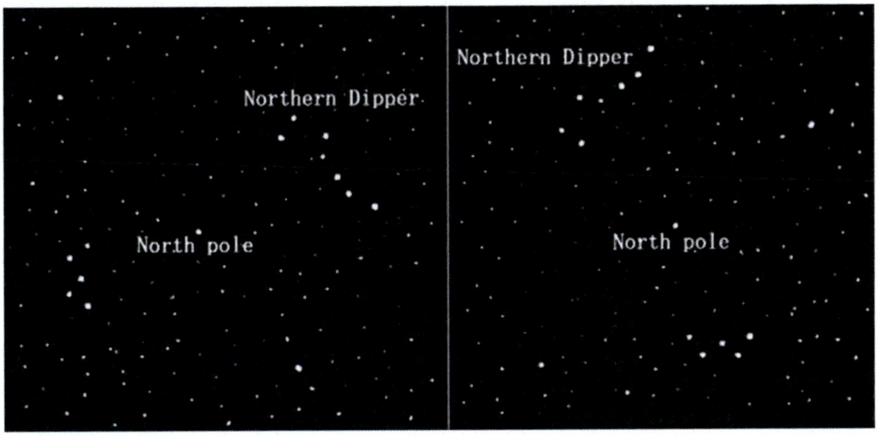

Fig. 9 Revolution of the northern dipper

outside this area will fall below the horizon for some time while revolving around the North Pole.

Standing at the same position at 21:00 the next day, we can see almost the same sky as we saw at the same time last night. However, the starry night has actually changed a little, which is difficult to detect unless we continue to observe it more. For instance, let's pick a bright star as a reference. If we see a bright star just on the top of a big tree at 21:00 on some evening, we'll find that the star is much higher above the treetop at 21:00 in the evening half a month later. The star will rise higher in a month and it will move to the western sky in three months. All these tell us that the rising time of stars is advancing day by day.

Before 1000 BC, the ancient Egyptians invented an astronomical instrument called Merkhet, which was specially used to observe the stars. It is a simple device. Place a plate with a slit in the middle on a column, and the observer changes his position to observe a star through the slit or waits for some star to pass behind the plate. It is used to measure the altitude of a star or the time when the star reaches its highest point of the day. Don't look down upon this primitive instrument—it could have been an "advanced instrument" in that remote ancient time.

We have understood through such careful observation that the starry sky varies at different moments on the same night because the stars are moving. And we have also known that the starry sky is different at the same time every night in a year because the stars rise earlier day by day. So, we see different stars at night in different seasons. We can see Cancer, Leo, Virgo at 21:00 in spring, Libra, Scorpio and Sagittarius at 21:00 in summer, Capricornus, Aquarius, Pisces at 21:00 in autumn, Aries, Taurus, Gemini at 21:00 in winter.

The countless bright or dim stars have their own places and the relative positions between one another remain constant in spite of the change of time, so they have been called fixed stars, which seem to be fixed in the sky. It's actually the whole starry sky that is rotating, though people say the stars change in positions. Perhaps this is why the ancient people felt they were at the center of the universe.

Humans divide the sky into many areas for better communication and identification of stars. They use imaginary lines to connect some brighter stars, which form beautiful patterns in imagination and are given interesting names. We can see the sky of countless and scattered stars in the schematic diagram of Fig. 10, when we face north at around 20:00 in early December in the middle latitudes of the northern hemisphere. Figure 11 shows the name of the northern sky in ancient China. With the Northern Pole as the center, what's around is called the Purple Forbidden Enclosure, symbolizing the imperial palace where the emperor, empress and the prince are supposed to dwell. On both sides of the Purple Forbidden Enclosure are the Supreme Palace Enclosure and the Heavenly Market Enclosure (outside the Figure), the former symbolizing the executive and the latter symbolizing the prosperous market. The Twenty-Eight Mansions encircle the Three Enclosures and they make a circle around the sky. Figure 12 informs us how ancient Babylonians and Greeks named the sky. They divided the sky into constellations of definite scope and gave them romantic mythological characters and stories, making the disorganized starry sky lively. The different dividing and naming of the sky also reflect the various cultures and thoughts of different nations.

What will be different when we move north or south to observe the sky in different places? The biggest difference is the height of the North Star. We can see the North Star is 45° above the ground in Harbin, 40° above the ground in Beijing, 32° above the ground in Nanjing, and only 23° above the ground in Guangzhou. In Harbin, the famous Northern Dipper can be seen at any time of the night. It is almost above us when it is at its highest while close to the northern horizon when at its lowest. But

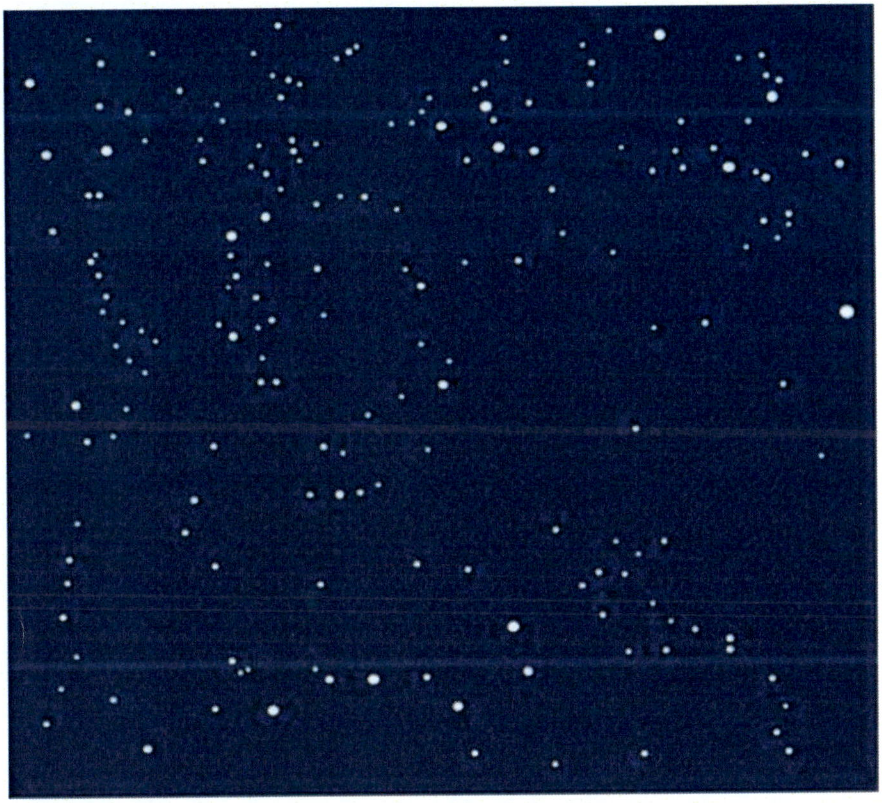

Fig. 10 Starry sky in arctic region

in Guangzhou, the Northern Dipper is in the northern sky when it revolves to the highest point, and it is invisible below the horizon when it is at the lowest point.

4 A Bright Moon

You may see the moon on a clear night. The moon must be much more attractive to ancient humans on cloudless nights without any light, though the moon is not as bright as the sun. The moon is a good friend of mankind because it is supposed to reduce people's fear of night with its bright and soft light and we can look at it for a long time. The shape of the moon is so beautiful and variable that human beings have praised it with poetic lines like "Before my bed a pool of light, I wonder if it's frost aground", "How rare the moon, so round and clear! With cup in hand, I ask of the blue sky", and "I only pray our life be long, and our souls together heavenward fly", as well as with charming lyrics like "The moon walks through the clouds which are like white lotuses," and "The sliver moonlight shines upon the boundless ocean."

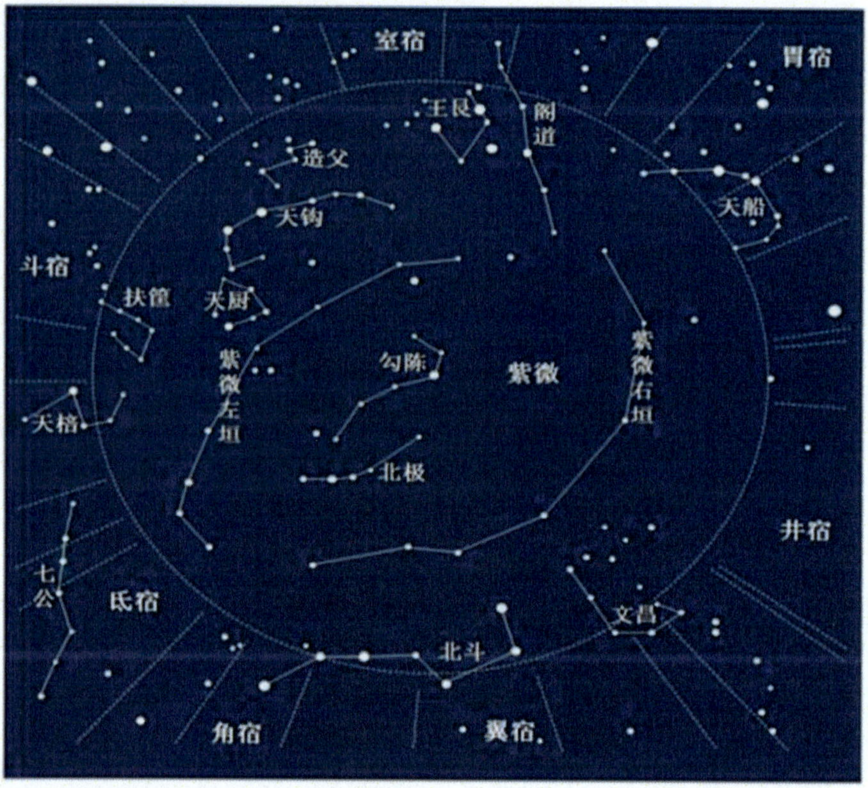

Fig. 11 The naming of stars in arctic region in Ancient China

Sometimes, we see such descriptions in the novel, "The sky turns dim as the sun sets, and a new moon rises from the east…" But it's a pity that this description is wrong despite its poeticity. However, many people have the same misconception that the moon, whether a new moon or a full moon, always rises from the east after dark and falls in the west near dawn the next day. As a matter of fact, we can never witness the slow rise of the curved crescent moon from the east because it only emerges in the western sky just after nightfall with a curved bulge pointing to the western horizon. Then it will follow the sun and sink into the horizon soon. As time goes by, the angular distance between the moon and the sun becomes larger and larger. And the crescent moon becomes a half moon gradually and finally a full moon day by day. Only when the moon is full, can people see it rising from the east by the time the sun goes down. The relationship between the lunar phase and sun-moon distance is shown in Fig. 13. The sun is setting into the western horizon in Fig. 13a. At the beginning of the month, the moon is close to the sun, and what we see then is a crescent moon that will follow the sun and sink into the horizon soon. When the moon is about 90° away from the sun, we see a half moon called the first quarter above our head at sunset. When the moon is 180° away from the sun, a full moon will rise from the east at

Fig. 12 The naming of stars in arctic region in Ancient Western countries

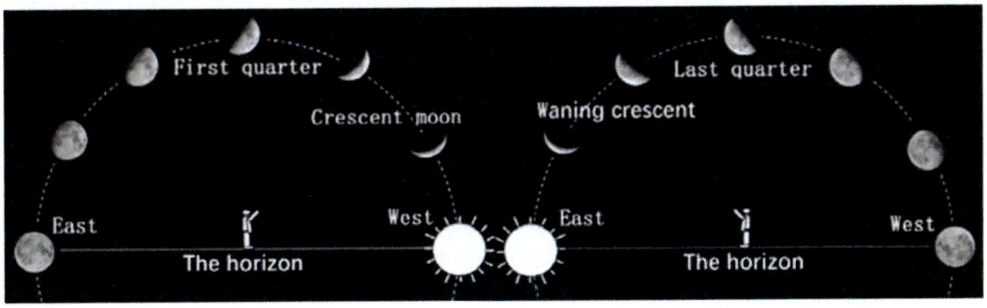

Fig. 13 Relationship between lunar phase and sun-moon distance

sunset and it will set from the west the next morning. As is shown in Fig. 13b, the sun is rising from the eastern horizon, and the moon, in another semicircle, comes closer and closer to the sun day by day, keeping changing its fascinating appearance. The full moon is to become the last quarter and then change into a waning crescent with a curved bulge pointing to the eastern horizon. The moon rises from the east before daybreak and will soon disappear in the brilliant light of the sun, so it is also impossible to see the waning moon set in the west slowly. One or two days after this happens, we cannot find where the moon is, and then we will see a new crescent moon when it appears again, in the western sky just after nightfall.

In general, the moon rises in the daytime in the first half of every month, so we can only see it sink in the west but cannot see it rise. We can see the moon rise in the second half of every month because the moon sets in the daytime, but we won't see the moonset. Only in the two or three days around the full moon can we see it rise and set, but it will disappear the whole night in the one or two days around the new moon.

There is a very stable cycle for the waxing and waning of the moon, which is the second unit of time provided by nature for human beings. The cycle is longer than the day, but not too long to be understood by human beings, and this appropriate cycle is called the "month". (Please note that the month mentioned afterward in this book refers to the "lunar month", that is the lunar phase of the waxing and waning of the moon, instead of the "solar month.").

Like the sun, the positions of moonrise and moonset also move within a certain range, but the change of the rising and falling points of the moon are faster and more obvious than that of the sun. It takes about 365 days for the sun to move back and forth, but only about 30 days for the moon to do it. People seldom pay much attention to how the rising point and falling point of the moon move due to their inability to see the rising and falling of the moon every day.

There are two famous lines in a popular love song in Xinjiang, China, which goes, "Half the moon climbs up, shining on my girl's dressing table." When does the "half moon climb up"? Fig. 13b tells us that the "half moon" is exactly the last quarter, which appears at around 24:00. The boy is so infatuated that he is still waiting for his beloved girl so late at night. Some knowledge of astronomy can therefore shed light on our understanding of the song.

Every day, the moon rises and reaches its highest point about 50 min later than it does the previous day (called "Upper Culmination" in astronomy), and it sets about 50 min later than the day before, so it seems that the moon moves more slowly than the sun. As we said before, the rising time of the fixed stars is advancing each day, hence the moon moves from east to west more slowly than the fixed stars. That is to say, with respect to the background of fixed stars, the moon moves from west to east.

Let's watch together at the time when it is easy to observe the moon. As is shown in Fig. 14, after sunset, a bright moon is hanging in the east when it is about to be the full moon. A bright fixed star (the bright cross in Fig. 14) lies not far east of the moon, and they rise together. The moon and the star, as they move over our heads and slowly westward together, are getting closer and closer, and then the moon will cover and pass the star. The moon, which is originally to the west of the star, has

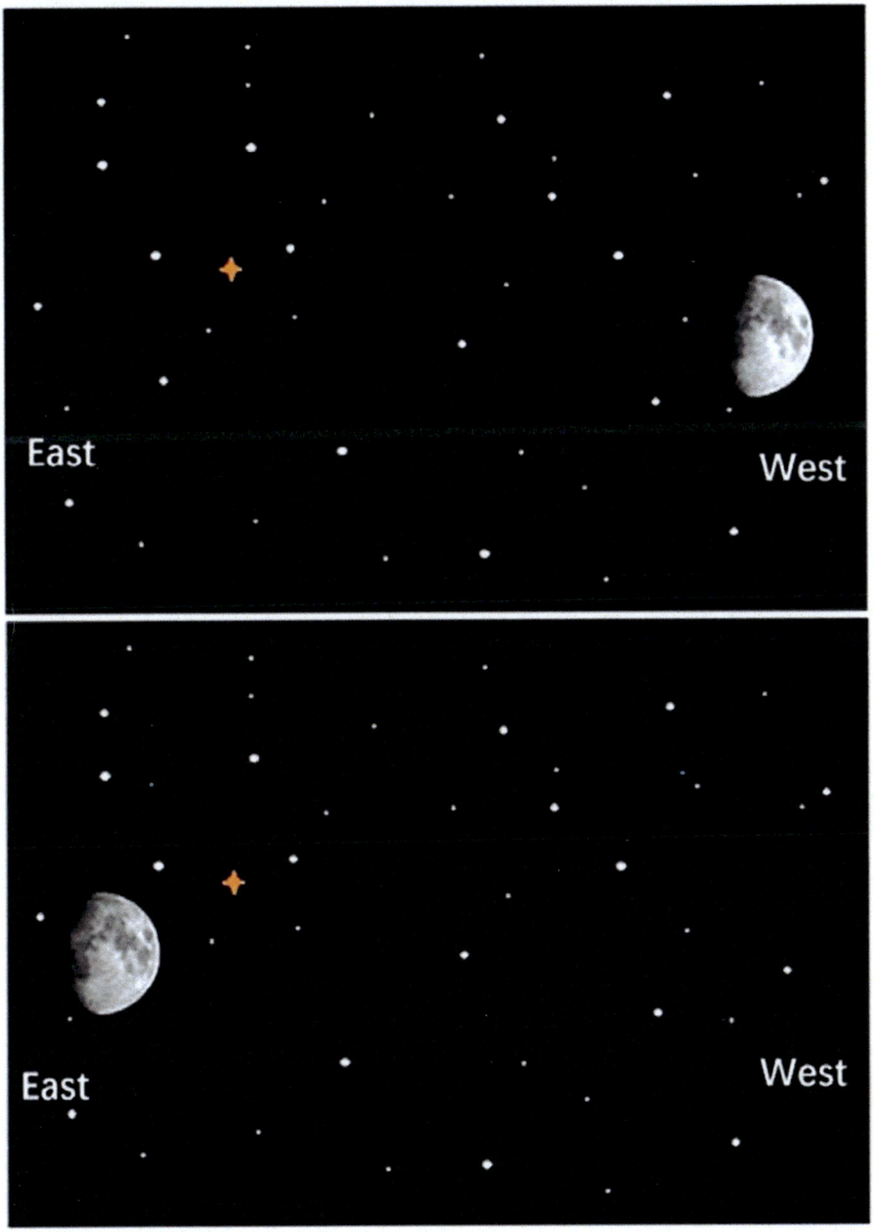

Fig. 14 Motion of the moon from west to east in the background of starry sky **a** early night on the eleventh day of the lunar calendar in a certain month **b** midnight on the eleventh day of the lunar calendar in a certain month

moved to the east of the star. Obviously, the moon moves slowly from west to east against the background of a multitude of stars.

It's not easy to observe the movement of the moon on a single night, but we can see that the moon moves from west to east against the background of stars, as long as we observe the moon at about the same time every night. For instance, we will see that the moon is in Libra at night on some eighth day of the lunar calendar in summer and in Scorpio the next night. For this reason, we can understand why the moon rises about 50 min later every day. But it is still a mystery why the moon rises in the east and falls in the west while moving in the opposite direction at the same time.

Please note that "the moon makes a circle in the sky", to be mentioned often hereafter, means that the moon moves from west to east against the stellar background and it is different from the apparent motion of the moon, rising in the east and falling in the west. The movement of celestial bodies from west to east is called prograde motion in astronomy.

The moon's movement under the background of stars has a fixed route, and we can roughly know it on the celestial sphere as long as we observe it for two or three months. We will find that the route of the moon and nearby stars form a close circle in the sky if we remember the stars near the route and their sequential order.

Sometimes, we can see the sun and the pale moon at the same time in the morning or afternoon. Of course, this can only happen when the moon is far from the sun and the sunlight is not so strong.

5 Unquiet Sky

The sky is not as quiet as we have described before, because you will see the meteor at times shoot across the night sky and disappear in the twinkle of an eye as long as you observe the sky for one or two hours. What kind of star is that? Will the sky consequently lose a star? These questions must have fascinated ancient people in the past.

With a careful search of the night sky with our naked eyes, we will find several stars different from numerous fixed stars. The fixed stars keep blinking, but these stars shine steadily and brightly. People must have initially thought that they belonged to the host of fixed stars, but it was found that they moved slowly from west to east in the background of the sky full of fixed stars, after a long period of careful observation. Sometimes they move faster, sometimes more slowly, sometimes they stop, turn around and move back for a short distance before they stop and move forward again, which looks very weird. The brightness of these stars changes periodically. People noticed these stars a long time ago and called them "Planets". There are altogether five planets visible to naked eyes, Mercury, Venus, Mars, Jupiter and Saturn. Figure 15 shows the actual position of Mars marked in the sky every seven days during its retrograde period, which enables us to clearly see how strange its trajectory is.

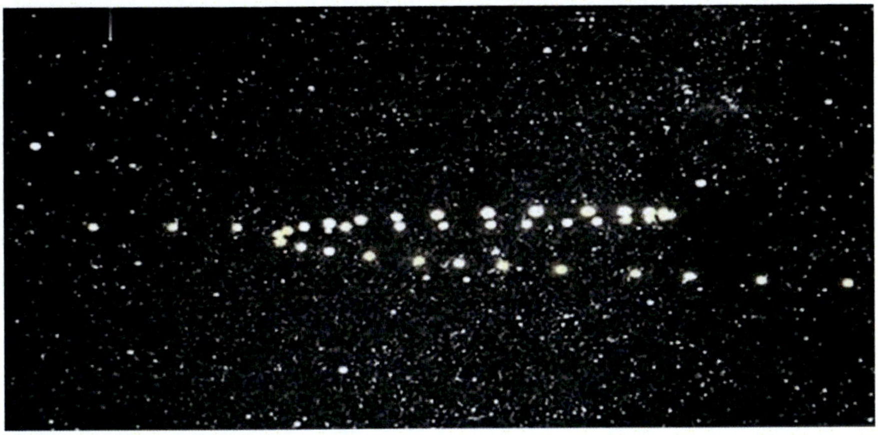

Fig. 15 Martian retrograde trajectory

After a long observation and record, you will find that Venus always moves back and forth on both sides of the sun, like a child leaning close to his mother. When Venus is on the east side of the sun, we will see it in the western sky at dusk and it was called evening star in the ancient time; when Venus is on the west side of the sun, we will see it in the eastern sky at dawn and it was called morning star in the ancient time. It is called east elongation in astronomy when Venus reaches the maximum angular distance at the east side of the sun, and it is called west elongation when Venus reaches the maximum angular distance at the west side of the sun. The angular distance between Venus and the sun never exceeds 48°. Venus moves past the sun when it travels back and forth from the east elongation to the west elongation and we cannot see it because of the strong sunlight. The situation is similar to that of Mercury but the angular distance between Mercury and the sun will not exceed 28°, and Mercury moves back and forth from one side of the sun to the other more frequently than Venus does.

The situations of Mars, Jupiter and Saturn are different, and there is no limit to the angular distance between them and the sun. When the planet such as Jupiter just rises in the east after the sun sets in the west, that is, when the angular distance between Jupiter and the sun is about 180°, we can see Jupiter all night as the brightest star during this period.

Venus and Mercury are always retrograde about the period they move past the sun when they change from evening stars to morning stars. Mars, Jupiter and Saturn are retrograde about the time when their angular distance from the sun is around 180°.

These strange "wanderers" will become the main topic later. It is their mysterious whereabouts that have stimulated the strong desire of astronomers to explore.

If you are lucky, you may see a star, with a tail pointing to the direction away from the sun, near the western horizon or eastern horizon at night or before dawn. That is the well-known comet, also called "broom star" by the Chinese because it is kind of like the broom used to sweep the floor at home. Where is the comet from and where will it go? What does its arrival mean? These questions were very mysterious to the ancient people.

The moon changes its look every day, especially the moon on the fifteenth day of the month, which is round, bright and perfect as a whole. But sometimes a special scene appears on this good day: a dark shadow emerging from nowhere covers the moon slowly and quietly. Sometimes it covers part of the lunar surface and then gradually fades away, and sometimes it covers all of the lunar surfaces and the moon turns bronze. The shadow will fade after some time and everything will come back again. These phenomena are called "partial lunar eclipse" and "total lunar eclipse". The lunar eclipse is visible to people who can see the moon anywhere when it occurs. The full moon is bright and we can look at it directly with our naked eyes and it is therefore easy to find if it loses a small piece. The chances of seeing lunar eclipses may be few but not rare.

Among the numerous astronomical phenomena, the solar eclipse is the most shocking. How horrible it was for ancient human beings if the brilliant sun suddenly lost a piece or even its whole. A round shadow covers the sun stealthily. It is not easy for people to be aware of it if only a small part of the sun is blocked. However, human beings, even animals, will panic when most of the sun is blocked and all is darkness. It will look like the end of the world if the whole sky is covered and the day becomes night when stars glitter here and there. These are called "total solar eclipse", definitely the most miraculous and shocking astronomical phenomena that human beings witness. It is still amazing for us to see a total solar eclipse, although human beings understand the causes of the solar eclipse and can predict it today. Besides the total solar eclipse, the annular solar eclipse is also one of the eclipses. An annular solar eclipse happens when the shadow covers the Sun's center, leaving the Sun's visible outer edges to form a golden ring around the shadow. Among all the eclipses, the partial solar eclipse is certainly the most common when only a part of the sun is invisible.

Overall, solar and lunar eclipses occur in almost equal numbers. The reason why people think the solar eclipse seldom happens is that they are unable to stare at the sun and it is therefore not easy for them to notice that only a small part of the sun is covered. Besides, the total solar eclipse occurs only in a small area. For instance, if we can see the total solar eclipse in Beijing, only the partial solar eclipse can be observed at the same time in Nanjing.

Then why do the solar and lunar eclipses happen? This was a great mystery for ancient people.

6 Eyes and Mind Combined

The basic astronomical phenomena we see with our naked eyes are described as above, and we have observed them carefully enough to summarize the universal laws of many astronomical phenomena. Today, it seems not difficult to find these phenomena and laws if people observe carefully. Nevertheless, it is very hard in fact! First of all, the Chinese idiom "The obvious usually goes unnoticed" precisely explains that many familiar phenomena will fail to draw our attention if we see them too often. For decades, some people have witnessed countless sunrises and sunsets, but they have never noticed that the rising and setting points of the sun change every day, and of course, they will not think about the why and how of the changes. Human beings were not born with the ability to analyze and compare, and not everyone is good at it.

Secondly, it requires a long period of patient, continuous and careful observations to notice the slow movement of the celestial body. For example, it is hard to see the moon moving slowly from west to east in the background of the celestial sphere in a short while, no matter how carefully you stare at it. It takes a long period of time for many astronomical phenomena to repeat themselves. For instance, the height of the sun at noon changes periodically and slowly, which usually requires long-term observation of tree shadows; and for some astronomical phenomena such as the revolution period of a planet, it takes several centuries of observation to accumulate enough data for comparison and analysis to find its law of occurrence. Besides observing with our eyes, we have to use our brilliant minds to think about it.

Therefore, it took so long for our ancestors just to get the above results of observation, and the previous description has condensed this process.

Today, has your curiosity been greatly aroused after we use our naked eyes to carefully observe and get familiar with the sky above us? Do you have any questions? And do you want to know the answers?

Of course, you may have known about it through textbooks or teachers' explanations, and the reasons behind a variety of astronomical phenomena are no secret to you. But have you ever wondered how we can find that the earth is spherical by observing only with our naked eyes and how we can conclude that the earth orbits the sun? If so, what kind of eyesight and intelligence do we need to do these?

The sky we see today is basically the same as the sky that was presented to our ancient ancestors. Anyone with normal vision can see it. Let's review how our predecessors thought about the sky! While following their thoughts, do not forget to ask ourselves: could we come up with the same idea when observing the same sky?

Chapter 2
The Motivation for Observing the Astronomical Phenomena at Night

After a long period of observation and recording, our ancestors gradually understood various celestial phenomena and their laws. Have you ever wondered whether it is necessary for human beings to observe the sky so carefully and what impels some people to spend night after night observing the sky, especially on those cold nights? There are three major motivations, urgent needs of human daily production and life, astrology (commonly known as fortune-telling) and the nature of human curiosity.

1 The Urgent Needs of Daily Production and Life

We cannot do anything without knowing time and place. For example, you must tell your friend when and where you'll see a movie, and he also needs to know the position of the seat. In the long evolution, human beings have learned to collect food, hunt, fish, raise livestock and farm. Where to do these? When to seed so that plants can survive and people can have good harvests? When do animals migrate and where is their destination? When does the river overflow? The answers to all these questions are inseparable from the prediction of the season and the determination of the location. The calendar is what people need to remember remote things, communicate and arrange long-term activities. As a result, determining the location and time became the most urgent problem for ancient people because it is closely related to our life.

(1) Journey of Location Determination

In the beginning, human beings had to determine the location by terrain and surface features. Remote mountains, rivers and even a big tree could be used as the reference of location in the era when the range of human activity was small. However, mountains and rivers would not meet the need with the increasing range of human activities and exchanges. For instance, when people in the south of the mountain and people in the north of the mountain made an appointment to go somewhere farther to do something, it would be easy to confuse them if they still used the mountain as

Y. Ge, *From Geocentric to Heliocentric: How Discoveries Are Made*,
https://doi.org/10.1007/978-981-16-7779-3_2

a reference. In the face of unbroken mountains and winding rivers, how could they set out together with a location as a reference? Something distant, visible, long-term and stable was needed as the reference. The sun, the moon and stars could exactly meet the needs. They are so remote that no one has ever been to where they rise or fall; they are obvious, especially the glorious sun, bringing light to the whole world; they always rise in one direction and fall in another on time, regular and completely predictable. Humans gradually learned to use astronomical phenomena to determine the location after long-term life experience.

This kind of utilization certainly had a development process from rough to precise. The rise and fall of the sun are undoubtedly the best reference during the day. In the beginning, people only generally called the direction of sunrise east, and the direction of sunset west. But the rising and falling point of the sun moved day by day. With more careful observation and the increasing needs of production and life, people learned to determine due east and due west according to the rising and falling position of the sun on specific days (vernal equinox and autumn equinox), or by measuring the two limit positions where the sun rises and falls, making their midpoint due east and due west.

After long observation, people also found that the sun at noon always pointed to the south no matter how the rising and falling points of it changed. So, they could distinguish the direction by observing the sun at noon.

The two methods mentioned above have great limitations. In real life, we usually need to accurately determine the direction at any place when we build a house or an altar. But it is impossible for people to determine the location only on the vernal equinox or autumn equinox, let alone to have a long observation for over one year just to do an ordinary thing. In addition, although the sun at noon is above our heads, ancient people lacked instruments that could accurately indicate the time and could not look at the sun directly. They could only judge the noon time by observing the shadow length of the pole, which changed little and was hard to be accurately measured at noon.

Later, smart people got the idea of drawing a big circle on flat ground by using the characteristic that the figure of the shadow sweeping over the flat ground is symmetrical in one day (Fig. 1). Set up a straight pole perpendicular to the ground in the center of the circle (point O), and the top of the pole's shadow falls on the circle which is marked as point A at some time in the morning. In the afternoon, there is always a time when the top of the pole's shadow falls on the circle and marks it as point B. A line is produced by connecting point A and point B points to the east and west. Make the midpoint (P) of the line (AB), and then draw a straight line between the center of the circle (point O) and the midpoint (point P). The straight line then points to due north and due south. This is the method adopted by the craftsmen in the Pre-Qin Dynasty (Paleolithic period to 221 BC), as is recorded in the *Kaogongji*, an ancient Chinese classic. Nicolaus Copernicus also mentioned this method in his *On the revolutions of the heavenly spheres* (*De Revolutionibus Orbium Coelestium*).

The pyramids built by ancient Egyptians are still standing after thousands of years. Each pyramid has an accurate position, which must be measured by using astronomical measurements in ancient times without the compass.

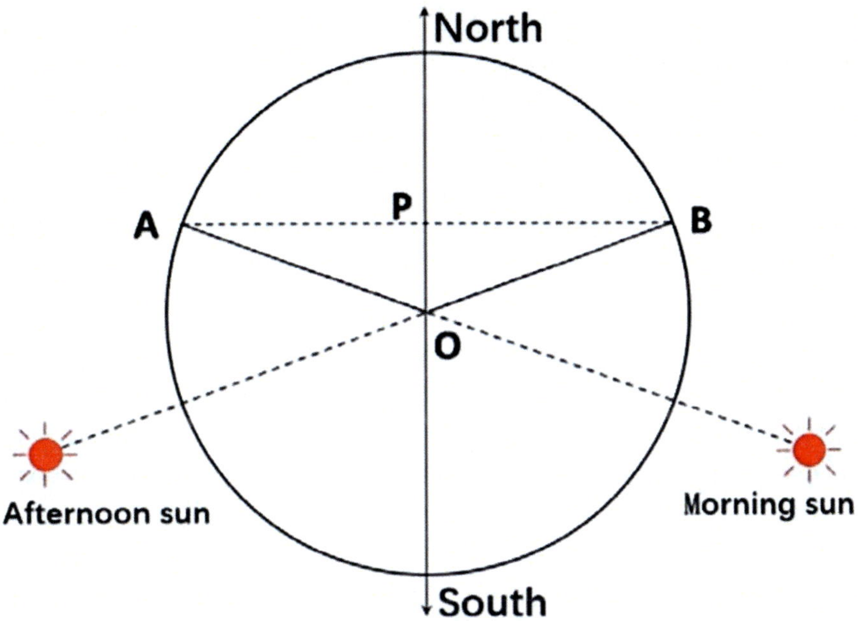

Fig. 1 Determination of position with a pole

The stars in the sky are endlessly revolving at night, but people find that a star is always in the center of the revolution and always fixed in the same position as long as it is a clear night, which is called a polar star. It's the best direction indicator mark at night. People call this direction due north and that star is called the North Star.

We know that it is not difficult to determine the other three directions as long as one direction is determined. We have no idea whether ancient people first determined the due east and due west according to the rising and falling of the sun, or the due south according to the sun at noon, or the due north according to the North Star. This process may have varied in different nations. It is a coincidence that no matter which method is first used, the other three directions obtained are consistent with those obtained by other methods. In other words, the line between due south determined by the sun at noon and due north determined by the north star is perpendicular to the line between due east and due west determined by the rising and falling of the sun, and the vertical intersection point of the two lines is the observers' position. It was hard to explain such a "coincidence" in ancient times when astronomical knowledge, cosmic theory and mathematical tools were still rudimentary.

(2) Name the Time

Human beings cannot work and live without the determination of spatial locations, as well as the division and positioning of time. It is easy to understand the division of time, but what is the positioning of time? The so-called positioning is to give a name to time, which is one of the important tasks for making a calendar. As time is continual and endless, human communication is almost impossible without the reasonable division and naming of time.

As we know, people not only need to remember the past but also need to predict and arrange the future. Suppose if one person asks another, "When were you born?" The man replied, "I was born 5326 days ago." After 20 days, he has to answer, "I was born 5346 days ago." Just imagine how troublesome it is! It is unimaginable to add one day to all the events every day and remember them in mind. No one can remember things like this.

If the time has a name, it is easy to remember the date of the event. For example, "October 1, 1949" and "10:00 on January 1, 2014" are the names of a particular time. Naming time is not as simple as naming a kid. Firstly, we need units of time, and secondly, we have to make clear the relationship between each time unit.

The sunrise and sunset, in addition to the alternation of day and night, provide mankind with the most basic unit of time—"day". Other units larger than it are supposed to be measured by it, while those smaller than a day are divided into equal parts according to people's needs.

What if nature just offered such a unit of time to humans? It would be very inconvenient, and it is not hard to imagine that humans have to create new and larger units of time out of necessity. But nature has been good to human beings because the changes of the moon have provided us with another unit—"month", whose length is perfect for calculation and memory.

How many days are there in a month? If we make careful and continuous measurement and calculation for a dozen successive months, we would know that a month has less than 30 days and more than 29 days. If we do it for a dozen years, that is, over 100 months, and take the average value, the answer will be much more accurate, which is enough for primitive people. Longer and more careful measurement and calculation are needed if we want to get more precise numbers. For example, what will happen if one continuously "count" for 4267 months? The result must be precise, right? Ancient Greek astronomers used ingenious methods to obtain an accurate average number of days of a month after "counting" for 4267 consecutive months, which will be mentioned later.

"Day" and "month" are still inadequate to measure longer time. So nature has provided a longer unit of time for human beings—"year". Compared with "day" and "month" whose definitions are quite clear, the "year" has an indistinct definition.

It is conceivable that primitive human beings must be sensitive to the change in temperature. Winter and summer are distinctly different in the middle latitudes of the northern hemisphere, where there are cycles of winter and summer, from severe cold to intense heat. Blossoms and deaths of plants, migration and reproduction of animals and regular floods are corresponding to temperature change. In spite of the long cycle of these changes, they were directly related to human survival, and their repetition greatly impressed people. As a result, people gradually realized this cycle and called it "year". It must be from the periodic change of temperature and phenology that humans first realized the unit of time—"year". The first men could ignore the long-term slow-motion changes of the sun, moon and stars, but they could by no means ignore the periodic changes of temperature and phenology, because they were directly concerned about the survival of human beings.

Theoretically, with the recognition of the cycle of year and three units of time including day, month and year as the basis, the formulation of the calendar and naming of time should be feasible. If a month was exactly equal to a certain number of days, and a year was equal to a certain number of months, then it would be simple to formulate a calendar and name the time. However, it was a pity that nature seemed to always play jokes on humankind. It had been long discovered the number of days in a month is between 29 and 30, but it was hard to get the precise number for a short while. It was more difficult to figure out how many days and how many months were in a year and the number was not integral anyway, bringing an enormous challenge for the formulation of the calendar. The year is vital to the season, harvest and human life, so it must be taken seriously. Accurately determining the length of a year and coordinating the relationship between year, month and day were the difficulties faced by all ancient civilizations when they formulated their calendars.

How long is a year? In other words, how many days are there in a year? The simplest and most direct way is to count. However, it is much more difficult to count the number of days in a year than to count them in a month according to the waxing and waning of the moon. Therefore, some ethnic groups with relatively slow development and evolution had never correctly counted the number of days in a year before the incursion of foreign culture. They only knew that a year had passed when they saw snow, green grass and ripe fruits again. In order to explore how many days there are in a year, human beings have carried out a long period of arduous work.

Which day should we start counting from? That is to say, which day should be set as the end of the old year and the beginning of the new one? This is a big difficulty. A large error will be made when calculating the length of the cycle if the beginning and the end of a cycle cannot be accurately determined. Men first came to know the unit of time "year" through periodic temperature changes, so the temperature would be a reference while counting the number of days in a year. For example, people can determine the number of days in a year by counting from cold days to hot and back to cold days, or from hot days to cold and back to hot days. But both hot days and cold days are long, spanning two or three or even four months, so which day of the year should be chosen as the starting point for counting? The coldest day or the hottest day? If so, how to determine which day is the hottest in summer or the coldest in winter? It is so difficult. Besides, how to unify the coldest day in two places that are a hundred miles or even a thousand miles apart? Perhaps some other methods can be adopted. It would be more feasible to take a certain place as the standard and count from the spring, such as the first day of peach blossom. Of course, there are many other methods. The ancient Egyptians may have counted from the day that the Nile began to flood.

In this way, we will find that the number of days varies from year to year, and sometimes it varies considerably. Why is that? Because the local fluctuation of temperature is large, and the natural phenomena corresponding to the temperature are also bound to have large fluctuations. So, what should we do? Is it necessary to precisely measure how many days there are in a month and how many days or how many months there are in a year? For ordinary people's daily life, we do not care much about the number of days in a year, and it does not matter if there is a day less or more in a single year. Sowing and harvesting are approximately the same in spite of the difference of a day. But many a little makes a mickle, as a calendar closely related to human life and in need of long use, it matters. If one day is counted less or more in a year, the months and seasons will be disordered after 30 or 50 years, and the astronomical phenomena on many specific dates will be misplaced. Therefore, the calendar needs the length of months and years to be as accurate as possible.

The calendar was a must for human life and production. People could not wait until they got accurate values to make the calendar. Instead, they had to make, test and adjust them at the same time. For instance, the day of peach blossoming might be chosen to count from at first, and then men could suppose that a year has 360 days when the peach blossomed again in 360 days. In the next few years, they felt nothing wrong about it, but as time went on, they would notice that the first day of a new year was getting farther and farther away from the day when peach blossomed. Hence the number 360 was slightly smaller than the days in a year. People had no idea how much smaller it was. They might as well adjust the number to 370. Again, they felt that it was fine in the next few years, but as time went on, the days of peach blossoming had passed for a long time before the start of a new year. It was obvious that the number 370 was slightly bigger, and human beings still had no idea how much bigger it was. But there was a rough boundary. The number of days in a year should be between 360 and 370. So, a certain number between them could be used to have a try.

It can be thought that humankind must initially calculate the number of days in a year by the method above. Therefore, the earliest calendar, which we call the phenological calendar, is an empirical calendar to determine the seasons through the cycle of long growth of animals and plants as well as the changes of summer and winter. In this way, we have to go through a period long enough (hundreds, even thousands of years) before we can make the number accurate between 365 and 366.

The calendar is a process of constant adjustment, testing and readjustment, which is still the same today. When does it need adjusting? First, it has to be adjusted when the accumulation of errors gradually appears and affects people's lives in the process of using it. Second, with progress and breakthroughs in theory and technology, the calendar is capable of being adjusted with more accurate data offered. In ancient times, it was discovered that the periodic movement of the sun revealed the changes of the four seasons, which was absolutely a great breakthrough in theory and technology of the calendar then.

We have no way to find out who first discovered that the change of seasons is actually caused by the periodic movement of the sun. In the prehistoric era before characters were invented, several major civilizations in the world had realized that the changes of seasons were closely related to the motion of the sun, though they had little communication with each other. But what we can be sure about is that this discovery must involve a long-term observation of the movement of the sun by ancient people, and it must have gone through an extremely long time. After finding the periodic changes of the sun's rising positions in the east and setting positions in the west, people would attempt to measure the length of the cycle, which would be more and more accurate with the development of the measurement method. Astronomers found that the cycle of the sun's rising positions in the east and setting positions in the west is about 365 days, which is consistent with the number of days obtained by observing the growth cycle of animals and plants as well as the changing period of summer and winter. The correlation between the solar movement and the changes of the four seasons has been confirmed consequently.

This discovery is great progress in human understanding of nature, because people have found the essence of the change of seasons. The temperature always keeps a specific relationship with the movement of the sun no matter how big its fluctuation is. The overall change of temperature is controlled by the sun, which moves regularly and stably. Compared with determining the number of days in a year by referring to temperature or natural phenomena closely related to temperature, observing the solar cycle to determine the length of a year is much more accurate and convenient.

In the Shang Dynasty (1600–1046 BC), the gnomon (Guibiao in Chinese) was already used to measure the shadow length of the sun at noon. It would be winter solstice when the shadow length was at its maximum, summer solstice when the shadow length was at its minimum, and the shadow length at vernal equinox or autumn equinox was in between. Therefore, it can be proved that Chinese people then had fully understood the relationship between "year" and the sun. It is a year from the winter solstice to the next one, or from the summer solstice to the next one. The length of the tropical year can be precisely measured with this method.

The gnomon is an astronomical instrument, combined with "Gui" and "Biao", for measuring the length of the shadow in ancient China, and its principle is very simple. The ruler or stone column perpendicular to the flat ground to measure the shadow of the sun is called Biao, and the slab with scale horizontally placed in the due south and due north is called Gui. The combination of the two can be used to measure the shadow length of the sun at noon every day. The photo of a gnomon at Purple Mountain Observatory is in Fig. 2.

It is more common to determine the summer solstice and winter solstice by measuring the shadow length with a pole than by observing the rising and falling points of the sun. Why? It is surely inconvenient to observe sunrise and sunset in mountainous areas. Even in plain areas, the chance of watching the whole process of the sunrise or sunset is much less than that of seeing the sun at noon. This is because there is more water vapor and dust near the land, and even on sunny days, it's often misty on the horizon when the sun rises or falls. Moreover, although the sun is not so dazzling in the morning or evening, it is inconvenient and difficult to measure the position of the sun with naked eyes or simple instruments. Measuring the shadow length with a pole, however, is more convenient and feasible, because the chance of seeing the sun at noon is greater and the strong sunlight at noon makes the shadow clear.

Besides, it is easier to determine the winter solstice than the summer solstice with the gnomon, as the shadow of a pole at noon on the summer solstice is short and its change is minimal around the summer solstice. For example, the shadow length of a 4-m-high pole, in two adjacent days around the summer solstice, has a change of just

Fig. 2 Gnomon

about 0.3 mm, while it has a change of shadow length of about 0.8 mm in the two adjacent days around the winter solstice, which is relatively easier to distinguish. As a result, the ancient Chinese paid more attention to the measurement of the shadow length of the pole on the winter solstice, and the starting point of "year" was therefore set on the winter solstice (Fig. 6 in Chap. 1).

Before the Western Zhou Dynasty in ancient China (around the tenth century BC), people had noticed that the starry sky was also related to the seasons. As is recorded in a Chinese ancient classic, it can be determined as spring if Hydra is seen in the due north of sky at dusk, as summer if Antares is seen, as fall if β Aquarii appears, and as winter if Taurus comes into our observation.

Or we can observe the Northern Dipper at dusk, which is bright in the northern sky. The pattern of seven stars is similar to the water ladle used in daily life. It is written in *He Guanzi* (an ancient classic of China) that "with the observation time fixed in the evening, the handle (formed by three stars of the Northern Dipper) points east on the vernal equinox, south on the summer solstice, west on the autumn equinox and north on the winter solstice." Fig. 3 is a schematic diagram combining the four directions that the Northern Dipper points. If we hold the Figure flat above our heads and compare it with the sky, there will be four directions presented in the figure: west on the left, east on the right, south above and north below.

By the Spring and Autumn Period and the Warring States Period (770–221 BC), the Chinese had already been able to make a calendar based on a year equal to 365 and 1/4 days.

The ancient Egyptians used to live on the banks of the Nile River, which swelled, flooded and covered the land it flowed through. The planting of crops could begin when the water receded, followed by the seasons of growth and harvest. The

Fig. 3 Four directions that the northern dipper points in four seasons

ancient Egyptians might have understood the cycle of "year" through this recurring phenomenon.

Sirius is the brightest star in the sky. Ancient Egyptians discovered that the Nile was about to flood on the day when the sun and Sirius rose almost simultaneously, that is, the day of the so-called Heliacal Rising of Sirius. So ancient Egyptians then used the Heliacal Rising of Sirius as a crucial basis for adjusting their calendar (religious rather than administrative), linking the length of "year" to the motion cycle of the sun in the starry sky.

Ancient Europeans also discovered the relationship between the seasons and the sun. Utilizing either natural or artificial signs, they measured and determined the length of the tropical year by observing the movement of the points of sunrise and sunset to determine the northernmost and southernmost points of the sun. For instance, Stonehenge, which was built in about 2000 BC or 3000 BC in Wiltshire, southern England, could do that according to researchers. Stonehenge is presented in Fig. 4.

The Hobbits in Arizona had an old tradition of starting a nine-day celebration each year when the sunset was near the southernmost winter solstice point. It indicates that they have already come to understand the relationship between "year" and the sun.

Fig. 4 Stonehenge (The picture is from the Internet)

However, it was much harder to understand "year" for those who lived in the ancient Middle East along the Euphrates and the Tigris rivers, where it was a tropical desert climate with no obvious seasonal changes. The two rivers, whose impacts formed Mesopotamia, did not follow the law of the annual ebb and flow of the Nile, so the ancient Mesopotamian people determined the length of "year" almost entirely by observing the stars. They made great contributions to astronomy, including 12 Zodiac Signs and the first use of the Sexagesimal system.

The length of "year" could be measured precisely through long-term observation of the starry sky and the motion cycle of the sun. With the increasingly accurate "month" and "year", the calendar also became more and more accurate after constant adjustment.

There was, of course, one more problem to solve. "Month" is not an integer multiple of "day", and "year" is not an integer multiple of "month" and "day", but we have to make the number of days in a month or a year an integer and the number of months in a year an integer. How could we do this? The only way was to make the number of days vary from month to month, and the same approach applies to the number of months and days in a year. Take the calendar used today as an instance, a lunar month has sometimes 29 days and sometimes 30 days, a lunar year has sometimes 12 months and sometimes 13 months, and likewise, a calendar year has sometimes 365 days and sometimes 366 days. But the final result of such an arrangement must ensure that, in the shortest possible period, the average number of days in a month, months in a year and days in a year are equal to the exact number of months and years. If people simply consider the relationship between the month and the day while ignoring the year (such as the Islamic calendar), or only that between the year and day while ignoring the month (such as the current international calendar), then the arrangement and coordination is relatively easier than that if the relationship among the three is considered (such as the Chinese calendar).

Human beings learned to adjust the relationship among "year", "month" and "day" in their early days. In the fourth century BC, the ancient Babylonians determined that there were seven leap years in nineteen years, that is, twelve out of nineteen years had twelve months a year, and another seven years had thirteen months a year. Chinese people before the Qin Dynasty (221–207 BC) already adopted the principle of setting seven leap months in 19 years.

There is no ultimate answer to the human measurements of month and year because the calculated result will be irrational numbers if we divide month by day or divide year by month or day. This could only be a process of making the result as accurate as possible, and a process that is infinitely close to the final answer. The calendar, in the same way, is the process of constant adjustment and improvement.

(3) Equally Dividing the Time

Production and life need not only units of time larger than "day" but also more precise units of time smaller than "day", which requires human beings themselves to divide the "day". As we can imagine, the initial division of time must be simple. A day can be divided into two parts according to the alternation of day and night, and then morning and afternoon, the first and second half of the night. What comes next? Continue dividing a day into 8, 16, even 32 and 64 equal parts? This problem certainly was puzzling to the ancients then. There were a 16-h system, a 12 two-hour system and a 100-quarter system in the history of China, among which the 12 two-hour system was the most well-known. In ancient Egypt, people divided day and night into 12 h each. In ancient India, a day was divided into 60 parts.

It is an interesting phenomenon that most ancient countries chose twelve or multiple of twelve while dividing a day equally. Why? We are not sure about the exact reason because there are no relevant records left, but some speculations are fairly reasonable.

A day means that the sun orbits around the earth once, and equally dividing a day seems intrinsically related to equally dividing the circumference. Easily drawn, a circle is the most common geometric figure. It is easy to separate the circumference into six identical parts with the radius of the circle, and six perfect equilateral triangles are made by linking six points from equally separating the circumference to its center. It is simple to separate a circle into 12 equal parts, which includes separating a circle into 4 equal parts, on the basis of the six equally divided circles. The angles of 90°, 60°, 45° and 30° are easily obtained by dividing the circle into four, six and twelve equal parts. All these angles play special roles in geometry. Therefore, it was not a coincidence that ancient people chose twelve or the multiple of twelve when they divided the "day" equally.

Time is invisible and intangible, so how can it be divided equally? Maybe someone will say: "Use the clock!" Don't laugh at this answer, because it is true that if you intend to divide the time, you have to use a clock, without which it is nothing but empty talk to divide the time equally. But the problem is that you have to invent a clock first, which is common today but was impossible in ancient times. Of course, there is still another method, that is, to observe the movement of celestial bodies. The sun, the moon and stars rise in the east and fall in the west and move evenly and regularly, and it is of great use to observe their motion to determine time because it takes almost the same time for them to move at the same angle. However, it was not convenient to observe on rainy days and nights, so people had to look for or invent things that change evenly, such as burning incense or the water dripping from a small hole. In the same vein, people invented the clepsydra and other timing tools that were used as the clock to divide time equally in ancient times.

After scientific equal division, more precise units of time, including the hour, minute and second were created.

But there was still another problem to be resolved. It is a continuous process, without a natural dividing line in the middle, between days, months and years, just like a circumference having no starting point and ending point. However, we need to distinguish yesterday, today, tomorrow and the day after tomorrow in real life, and what needs to be clear is what time the month or year begins and ends. Among them, the key point is to demarcate the day. So when does a new day start from? People work and carry out production in the daytime, so humans must initially regard the dawn as the start of a new day. Even today, people still regard getting up in the morning as the beginning of a new day in terms of psychology and living habits. But what time exactly? It is dawn at 7:00 in winter in Beijing, so how about setting seven o'clock in the morning as the beginning? But in summer, the daybreak is at 4:00. Can the next three hours be counted as yesterday? If so, how awkward it is. Likewise, it is awkward if four o'clock in the morning is set as the dividing line in winter.

The sun rises at different times every day, so there is no suitable time in the morning for demarcating the day. Through long observation, astronomers found that the sun reaches above our heads precisely at noon whether it is summer or winter, which provided a good reference point for the division between day and day. However, it will bring great inconvenience to real life if such a division comes into being, which is not feasible. So, astronomers chose midnight corresponding to noon as the end of the day and the beginning of a new day. Today, this division is universally adopted around the world.

With the location of time and space, the position and the calendar, and the name of time that is easy to remember, it is possible for humans to communicate in a broader scope and carry out various complex collective activities, such as gathering and storing fruits, hunting, fishing, animal husbandry, farming, large-scale construction projects and wars. The development and progress of society require more and more accurate calendars, which impels astronomers to observe and analyze the sun and the sky carefully. Measuring the celestial bodies is a premise for formulating a calendar, and the quality of the calendar has to be tested by the measurement. The urgent needs of human daily production and life promote the continuous development of astronomical research.

2 Comprehending God's Will

(1) The Legend of Astrology

The vast sky is full of mystery for ancient people, wind, rain, thunder and lightning in the sky, sun, moon and stars also in the sky. The astonishment, fear and awe of unusual celestial phenomena, especially eclipses of the moon and the sun, haunted them, and without any scientific explanation, they felt or believed almost instinctively that the heavens were where the gods dwell. Various strange celestial phenomena were closely associated with all kinds of phenomena of human society, and they were omens for human beings, countries and even individuals. Humans instinctively hoped to predict and master their destiny, or at least to avoid misfortunes.

In ancient times when science and technology were at a rather low level, limited by knowledge and thoughts, few people could get rid of the shackles of gods, fate and other concepts. So, observing the sky at night, "comprehending" the will of God, and predicting destiny became a kind of special knowledge and occupation. For example, there was a kind of Horoscope in Babylon, at least six of which still survive today. The oldest one of them is from 410 BC, which records the celestial phenomena at one's birth. The moon is below the Scorpio; Jupiter is in the Pisces; Venus is in Taurus; Saturn is in the Cancer; Mars is in the Gemini; Mercury is invisible.

Astrologers have basic judgments for all kinds of celestial phenomena. For example, those who were born when the moon rises from the horizon will have a bright and happy life, and steady growth and longevity; those born when Mars rises above the horizon will be injured in childhood and die of diseases. The kid will be lucky if Jupiter rises from the horizon and Mars falls below the horizon at his birth, but his wife will die before him, and so on. Another instance is that ancient Chinese people gave a complete set of astrological significance to the five planets when they move through or stay at the Twenty-Eight Mansions and other asterisms. If Mars moves to the Antares when something called "stationary" in astronomy happens, then it is called "Mars Staying at Antares", which is regarded as a great boding of evil in Chinese astrological theory and has explanations such as the death of an emperor and the chaos in the whole world.

We have no idea how the ancient astrologers connected the future of the world with a certain celestial phenomenon, but we can imagine that perhaps there was some "reason" though it was a superstition. For example, in the beginning, when a tribal leader died shortly around a solar eclipse, others might associate this with the celestial phenomenon. They wrongly believed that there was a causal relationship, and thus recorded it, which became the "judgment". Of course, the astrologers' far-fetched explanation and "flexible use" also played an important role.

The main purpose of astrology was to predict the fate of nations and ethnic groups, such as natural disasters, a good or poor harvest of animal husbandry and fishing, the outcome of the war, the safety of the emperor, etc. There were also ordinary people believing astrology, which would determine whether the building of houses, weddings and funerals could be carried out. A lot of miraculous legends of astrologers circulated in all countries of the world.

As is written in the *Romance of Three Kingdoms,* "That night Zhuge Liang was carried into the open and gazed up at the sky. 'That is my star,' said he, pointing to one that seemed to be losing its brilliancy and to be tottering in its place." Sima Yi, Zhuge Liang's old rival, also understood the celestial phenomena, "Sima Yi watched the skies. One night, a large red star with bright rays passed from the northeast to the southwest and dropped over the camps of Shu. It dipped thrice and rose again. Sima Yi also heard a low rumbling in the distance. He was pleased and excited, and said to those about him, 'Zhuge Liang is dead!' This notion of "When a star falls, a soul ascends to God." is just from ancient astrology. Ever since, people have spoken of the death of the great and famous as "the fall of a Giant star".

Such a story is told in the *Book of Jin·Zhang Hua.* In the early Western Jin Dynasty (265–317 AD) before the State of Wu (229–280 AD) perished, the Purple often appeared between the Dipper Mansion and the Ox Mansion, and it was regarded as the sign of the State of Wu being still powerful. Because the two mansions corresponded to the States of Wu and Yue according to the division of stars in ancient China. But Zhang Hua, a senior official of Western Jin, insisted on conquering Wu country regardless of the celestial phenomenon. The Wu State was destroyed before long, but the Purple turned brighter, which proved that the Purple was not related to the national power of Wu country, so what was the exact reason? Zhang Hua heard that Lei Huan from the city of Yuzhang was skillful at astrology and asked him about it secretly. Lei

Huan replied, "That is the soul of swords rising in the sky", and the swords were in Fengcheng county, Yuzhang city. In a high position, Zhang Hua managed to appoint Lei Huan as magistrate of Fengcheng. After he took office, Lei Huan immediately dug more than 40 feet deep under the foundation of the county prison and got two well-known swords in ancient legends—"Longquan Sword" and "Taie Sword". That night, the Purple between the Dipper Mansion and the Ox Mansion disappeared.

Nanjing is an ancient city in China, also known as Jinling. Legend has it that in the Qin Dynasty, someone suggested to Qin Shi Huang that there was an indication of the king in the corresponding southeast direction according to the celestial phenomena, which was a great threat to the Qin Dynasty. So, Qin Shi Huang asked to make a small golden man to be buried in Nanjing somewhere underground and announced that "it is not in front of the mountain, not behind the mountain, not in the south of the mountain, not in the north of the mountain. Whoever gets it will make a rich country." In this way, the gold seekers were lured to dig deep around the mountain, cutting off the mountain, destroying the *Fengshui* landscape and the potential threat of Nanjing. Therefore, Nanjing is also known as Jinling (Jin means "gold" in Chinese).

There is a river called the Qinhuai River in Nanjing. It is said that Qin Shi Huang was always thinking about how to make the kingdom last forever after unifying China. Someone reported to him, "Observing the sky at night, I saw an indication of the king appearing in Jinling, so there must be a threat to our nation, Your Majesty." "Oh, really?" "A pleb like me does not dare to lie to you. A king will be born there after 500 years." Qin Shi Huang asked, "Do you have any solution?" "A river can be dug here to cut off this *Fengshui* landscape and eliminate the threat." Therefore, the emperor ordered to dig a river in Nanjing, which is called Qinhuai River because it was built by Qin Shi Huang.

These are simply two legends, but coincidentally, all the dynasties in Chinese history that set their capital in Nanjing were short-lived. Wu country in Three Kingdoms Period, Eastern Jin Dynasty (317–420 AD), Song country, Qi country, Liang country and Chen country of the Southern Dynasty were all small dynasties of short duration. Zhu Yuanzhang, emperor of the Ming Dynasty, established the capital in Nanjing. Zhu Yunwen, the grandson of the emperor, ascended to the throne after Zhu Yuanzhang's death, and he is also known as Jianwen Emperor, with the reign title as Jianwen. But soon after, he was overthrown by his uncle Zhu Di, who moved the capital to Beijing, so the Ming Dynasty in Nanjing was also short-lived. The Taiping Heavenly Kingdom and the Kuomintang government both set the capitals in Nanjing for a short time.

Many legends about astrology also spread in Europe.

Nero Claudius Drusus Germanicus, cruel and notorious, was a famous tyrant in ancient Rome (753 BC–1453 AD). His mother Julia Agrippina Minor asked an astrologer to do fortune telling for her son Nero, and got the reply that Nero would be the emperor, but he would kill his mother when he mounted the throne! Agrippina said instantly, "It is all right if only my son can become the emperor." She was really killed under the order of Nero later.

According to an astrologer's prediction, the Roman emperor Domitian would be killed by swords in the future, which was followed by a more explicit prediction: the emperor would die on September 18, 96. Domitian was very superstitious originally, and he had been in a constant panic ever since. He gathered guards on 17th to strengthen the guard, waiting for the moment that "the moon moves to Aquarius, stained with blood". At the same time, the emperor took the prediction to his heart and ordered the astrologer, who spread the prediction, to come and asked if he could know how he would die. The astrologer replied that he would be torn to pieces by dogs. The emperor wanted him to die in a different way, so he ordered the astrologer to be executed and burned his body with fire. However, a sudden thunderstorm put out the fire when he was being burned and a group of wild dogs came to eat the body. On the 18th, it seemed a year passed for Domitian. How he wished time could fly! He asked his servant to check whether it was the time again and again. Finally, the servant was so annoyed that he reported it had already passed the time. The emperor was relieved and went to take a bath. At this time, a conspirator who was planning to murder the emperor asked if he would like to listen to him reading something in the bath (it was a popular practice in Roman aristocratic stratum to listen to people reading aloud while having a rest), and the emperor accepted it. But it turned out that the conspirator pulled out a short sword and stabbed it into the emperor's chest in the bathroom. At last, Domitian was killed by the sword.

The queen of King Henri II of France once, under a pseudonym, asked an astrologer to tell the fortune of her husband. The astrologer predicted that this man would die in a duel. The queen laughed and thought, "Who dares to challenge the king or ask for a duel?" Later, during a contest for his daughter's wedding celebration, King Henri II was stabbed by the chief of the Scottish guard. The spear stabbed through the crack of his visor (the warriors were covered from head to foot with iron armor then), and his head was injured. Ten days later, the king died at the age of 40.

These legends and stories about astrologers, whether they are true or false, or coincidental, all show that ancient emperors attached great importance to astrology.

(2) Contribution of the Astrologers

In the eyes of most of us, astrology is fundamentally different from astronomy: the former is superstition, while the latter is science. Superstitious thoughts and behaviors will only hinder the development of science, so how can they become the driving force of astronomy?

The "ancient astronomers" we often mentioned include many astrologers. The reason why we call them astronomers is that astrologers were studying astronomy, though their main purpose of observing celestial phenomena was to predict the future rather than to find out the reason behind them. At least they had to observe, measure and record all kinds of astronomical phenomena all the time and study their laws, which were essential parts of their astrological activities. For example, without astronomical knowledge, it was impossible to make the records in the Horoscope mentioned above or to predict the astronomical phenomena at a certain time in the future. Whether some celestial phenomenon occurred as predicted was also the standard to test the astrologer's ability.

Li Chunfeng, a great astrologer in the Tang Dynasty of China (618–907 AD), predicted that there would be a solar eclipse someday. Emperor Taizong of Tang was not happy after hearing it because the eclipse was a boding of evil to the king, so he asked Li Chunfeng what if it would not happen, and Li Chunfeng said that he could be guilty of death if so. When the day came, Emperor Taizong waited with Li Chunfeng in the palace courtyard, but the eclipse did not happen after a long time which made the emperor think the forecast was wrong. Then he said to Li Chunfeng, "I allow you to go home and say goodbye to your wife and children." But Li Chunfeng did not panic and said, "It is not the time yet." Then he drew a line on the wall and told the emperor that the eclipse would happen when the shadow reached that line. It turned out that his prediction was completely right.

It can also be said that the purpose of astrology is superstitious, but its foundation must be scientific, and methods of observation and research must be scientific. In the era before modern astronomy came into being, the need for astrology was the most important reason why the ancients constantly recorded a large number of astronomical observations year after year.

Astronomy and astronomer are, to a large extent, a pair of modern concepts. In ancient times, there were only astrology and astrologers. Very few of them observed celestial phenomena not only to predict good or evil, but also to find out the real reasons behind them, such as Claudius Ptolemy (about 90–168 AD) in ancient Greece and Zhang Heng in ancient China, etc.

In ancient times, astrology laid the foundation for the development of astronomy and accumulated astronomical knowledge, in both the western and eastern worlds. Most of the early works of astronomy contained astrological elements, and astrologers and astronomers were still indistinguishable until the Renaissance and two or three centuries after in Europe.

Astrology had left behind abundant astronomical observation records for later generations. Some astronomical laws with a long period of change were discovered by later generations in the analysis of the detailed observation records of predecessors. For example, there was a group of astrologers in ancient Babylon who specialized in observing the sky and kept tracking any changes in the sky. When they found an unusual celestial phenomenon, they would interpret it as a forewarning of disaster given by God to the king and the people. After continuous observation and record for seven centuries, a series of over 70 astrological lists containing about 7000 omens came into being. Scrupulous scientists gradually found that some astronomical phenomena that they thought accidental repeatedly appear. By analyzing such repetitive phenomena, they gradually identified and determined the cycles, or the laws of the movements of the sun, moon and planets.

The great majority of the abundant astronomical phenomena in ancient China were recorded for astrological purposes, and they have become precious historical heritage and are of great value in solving some major astronomical problems. The long-standing observation records are particularly crucial to many aspects of astronomical research, because the observed objects vary greatly in time span, far from being comparable to individual life. The research of later generations all resorts to ancient observational records.

Also, the exact location and time were needed for numerous religious activities, such as determining the orientation of the temple construction, the date and time of important religious rituals and sacrificial activities, which could not be done without astronomical knowledge.

If human beings do not need "fortune-telling", the development of astronomy is likely to be much slower, right? Therefore, this great need for social psychology for humans is also an important driving force to promote astronomical research and development.

Even today, in the twenty-first century, we can still see the vestiges of astrology. A lot of young people know their own horoscopes and some are very interested in the "fortune" of their horoscopes. The idea that a person's destiny and character can be predicted by the celestial phenomena at the time of his birth comes from ancient times. Of course, most young people just think it's fun and do not take it seriously.

3 Just Out of Curiosity

Determining locations, predicting seasons, formulating calendars and telling fortunes are all practical purposes needed in people's real-life, powerfully promoting the development of astronomy. However, if we do everything for practical purposes, humankind would not think over anything seemingly useless at that time or seek the truth behind the phenomena. Astronomy would stop advancing and have no breakthroughs. So the third main driving force for astronomy to move forward is the innate curiosity of human beings. Curiosity is the main cause of scientific knowledge, especially in astronomy, and the exploration of the structure of the universe is mainly due to human curiosity.

The vast and mysterious sky not only makes us wonder, tremble and awe of it, but also makes us extremely curious about it. How high is the sky? How vast is the land? What is in the sky? Is the ground square or round? What is underground? Is the sun a fireball? Why does the moon change its shape? What is on the moon? Why are there eclipses of the moon and the sun? How do the sun, moon, stars get through under us? We have no idea about the reasons, but really want to know why. It is human nature to ask and think out of curiosity and to explore due to incomprehension.

Aristotle, the ancient Greek philosopher, said, "All men by nature desire to know." People's exploration of philosophy from ancient times to the present originates from the surprise at all things in nature. At first, people were shocked by all kinds of confusing phenomena and then accumulated explanations little by little. Gradually, they were able to explain some major issues such as the movement of the sun, moon and stars as well as the birth of the universe. Confused and astonished while facing the unknown, they were ashamed of their stupidity. The only purpose of exploring philosophy for them was to get rid of stupidity. They engaged in academic research for the sake of knowledge, not for any practical purpose.

How high is the sky? Is the ground square or round? How do the sun, the moon and stars pass through the ground? Would it affect ancient human life whether they knew the answers to these questions? Even over 500 years ago, what did life have to do with exploring whether the sun revolved around the earth or the earth revolved around the sun? Similarly, today, whether the universe was born in a big bang or it has no beginning and end at all? Is there a black hole in the universe? The study on such questions seems to be of no practical use to us now, maybe the same in the future. But curiosity pushes people to try to understand all of these and to explore the unknown world.

Famous scientists at all times and in all countries have contributed a lot in a certain field with efforts, mostly because of curiosity and interest. Astronomers' hard pursuit of the structure of the universe generation after generation can perfectly prove this.

In the speech at the celebration of the 60th birthday of his beloved elder, Planck, Albert Einstein, a famous scientist, said that many people love science because it gives them extraordinary intellectual pleasure, and science is a special entertainment for them.

On 21st, August 2002, at the Chinese Youth Mathematics Forum, Chen Xingshen, a famous 92-year-old Chinese mathematician, wrote an inscription: "Fun Mathematics" (Fig. 5). Many times when he was asked why he studied math all his life, he replied, "It is fun."

Fig. 5 Inscription of Chen Xingshen (The picture is from the Internet)

Ding Zhaozhong, an American Chinese physicist and Nobel Laurant, led the Alpha Magnetic Spectrometer (AMS) experiment project to explore the dark matter and antimatter of the universe in international space. When someone asked him, "What is the use of the invisible dark matter? Why do you study the origin of the universe? Does it matter to the world and life?" Ding Zhaozhong answered, "I do not know what I can do with my research. I do my work just to satisfy my curiosity."

Human curiosity is the motive force of science, and human beings need to "think impractically". Curiosity is the nature of humans and comprehension is the pleasure of life. With curiosity, leisure and freedom, we can "think impractically", without which there will be no "fantastic thinking". It is of great significance for the development and progress of our country to stimulate young people's curiosity and encourage them to "think impractically and innovatively".

Like many other disciplines, astronomy also makes progress step by step in the actual need of production, life and society, and is further driven by curiosity.

Chapter 3
Mysterious Whereabouts of the Sun, Moon and Stars

Imagine if you were a person living in ancient times, what would you think while carefully watching the sun, moon and stars every day? Were you shocked, curious, or wondering why?

The sky is so vast and mysterious, and the easiest way is to attribute all of these to gods. But we have learned that even in the ancient primitive times, there were still people pursuing truth persistently. Some relatively simple questions had more reasonable answers after countless wise men had thought over and analyzed them carefully. In history, ancient civilizations had reached some basic inferences successively in spite of little communication between them, and these conclusions are crucial to the subsequent research.

Readers need to know that although the first two chapters are easy to understand, the following explanation requires you to think and give full play to your imagination, so that you can keep up with the pace of ancient astronomers to pursue the truth of the world and to explore the scientific truth with them.

1 Where Do the Stars Hide?

Every morning, as dawn breaks and the sky gradually lights up, the bright stars gradually dim and disappear. When the afterglow of the sun fades away and the night slowly falls, the dazzling stars will appear in the sky one by one. Ancient people must wonder where the stars are hiding in the daytime. If someone announced to everyone that stars also exist during the day, they would be very surprised, right?

Who found that there were stars in the daytime? And when? I'm afraid we will never know the answer, because human beings had known that there were stars in the daytime before the record of writing, which is a really amazing judgment in ancient times.

The judgment must come from long-term observation of the stars. After ancient astronomers (or called astrologers) became familiar with all the visible stars in the sky, some would think that the bright stars in the sky, which always kept a fixed position

Y. Ge, *From Geocentric to Heliocentric: How Discoveries Are Made*,
https://doi.org/10.1007/978-981-16-7779-3_3

and revolved at the same uniform angular speed, could not suddenly disappear when the day came and suddenly return to their respective position at night. Every morning, as the sky gets brighter, the stars do not disappear in the blink of an eye, but gradually fade away. Every evening, with the fall of night, the stars do not appear suddenly, but slowly become visible. Compared with the sun, their brightness is so weak that they are covered by the radiant sunlight, just like the campfire in the distance that is almost invisible in the daytime.

Therefore, people had reasons to believe that the stars should still be in the sky during daytime, and they were moving slowly from east to west at the same speed as they were at night. But the stars and their movement were invisible because of the bright sun.

In winter, the day is shorter and the night is longer, so it can be found that the stars rising from the east shortly before dawn will appear near the horizon of due west when darkness falls, which is exactly the place it should reach after moving for a day. In the same way, the stars in due north that never fall on the horizon fade away when the dawn comes. And they will gradually appear again at night, whose positions have changed. All of them have revolved around the North Pole about half a circle during the time when people cannot see them. All these show that whether it is day or night, the stars revolve along their respective orbits in their respective positions and never stop. What is more convincing is that in the short period when the total solar eclipse happens, the sky is dark as night and people see the stars in the sky with their own eyes, and they are the stars that are supposed to reach those positions at that time according to calculation, which are just near the sun! This directly proves that there are stars in the daytime.

Also in winter, it can be found that the stars that slowly fall into the western horizon at night will rise from the eastern horizon before dawn and maintain their original relative position. What are the stars hidden underground like? How do they get through under the ground and keep their original order? The sun and the moon set in the west and rise in the east, but how do they get through under the ground? It is hard for people to imagine and understand.

Incredible as it is, ancient people who paid attention to the sky realized through long observation that every star has its own fixed position, and they are still in their respective positions whether in the daytime or when they fall below the western horizon.

Astronomers have been familiar with the stars visible to the naked eye, and even some ordinary people were able to recognize the main bright stars. Countless stars formed an almost complete enclosed sphere, so the concept of a complete celestial sphere came into being in some astronomers' minds. When we see the stars in half of the celestial sphere, the stars in the other half should be under the earth. The sky above the earth and the sky below the earth form a complete sphere. The celestial sphere, as a whole, rotates around an invisible axis which is the line between we observers and the North Pole. The sky overhead moves from east to west and the sky underfoot moves from west to east, so the stars falling under the western horizon will rise again from the eastern horizon. Thanks to this rotation, we can see more than half of the stars in the celestial sphere in a whole night, and all the stars in the almost complete spherical sky will be seen after continuous observation for months. Figure 1 shows the diagram of the spherical sky.

The rotating celestial sphere certainly has two corresponding poles. One is the north celestial pole that we can see at night, and the other is the point where the line between the north celestial pole and we observers extends to the other direction and intersects with the celestial sphere. This point must exist in the south below the horizon though we cannot see it, which is called the south celestial pole by astronomers.

Therefore, observed in places such as ancient China, Greece, Babylon, India, the stars in the area around the north celestial pole will never fall into the horizon, which is called the circle of perpetual apparition, while the stars in the area around the south celestial pole will always hide under the horizon, which is called the circle of perpetual occultation. These can be clearly seen from the schematic diagram (Fig. 1). The reason why "almost complete" is used in "formed an almost complete enclosed

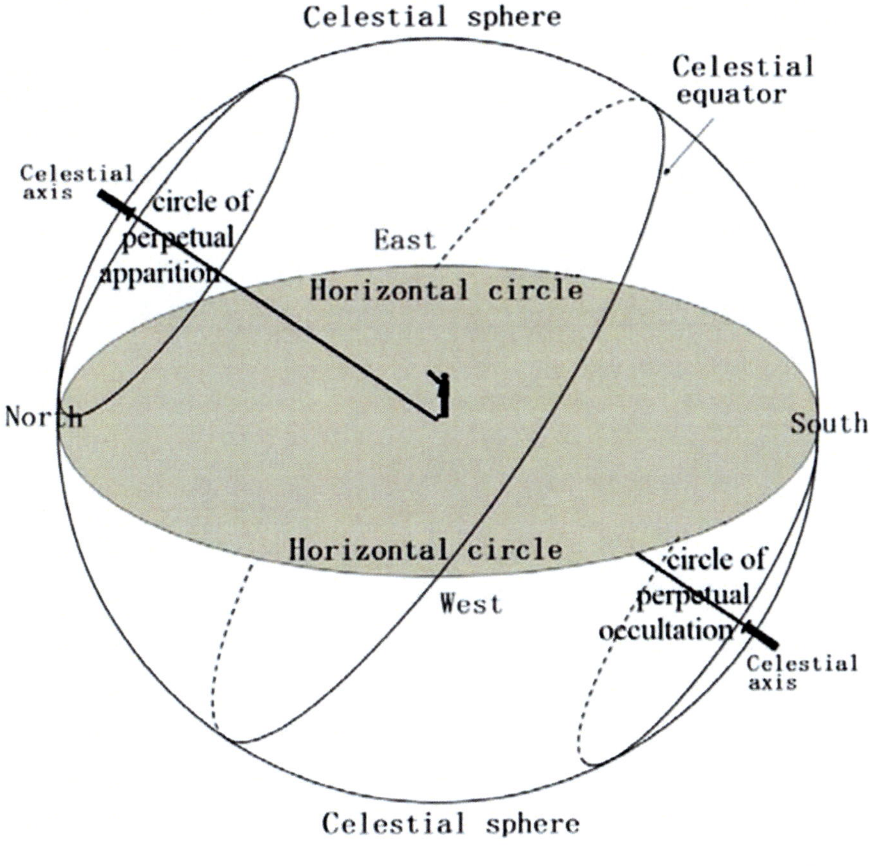

Fig. 1 Spherical sky

sphere" and "all the stars in the almost complete spherical sky will be seen after continuous observation for months" mentioned before is that the sphere does not contain a region of stars around the southern pole. Because the axis of the celestial sphere is tilted to the horizon, we can thus understand why the arc of the sun, moon, and stars from rising in the east to setting in the west always appears tilted to the south.

In the sixth and seventh centuries BC, the Anaximander of ancient Greece realized that the universe was not a plane or a hemisphere, but a sphere. He believed that the sky revolves around the north star, and the visible dome of the sky is half of a complete sphere. The flat disk-shaped earth is at the center of the sphere, surrounded by air, stars, the moon, planets and the sun.

In the 4th century BC, Shen Dao of the Warring States period in ancient China believed that the sky was like a ball, and its axis of rotation was tilted to the horizon. This view became clearer in the Han Dynasty. Zhang Heng (78–139 AD) of the Eastern Han Dynasty said in his *Annotation of the Armillary Sphere* that the circumference of the circular sky could be divided into 365 and 1/4° (In ancient China, the circumference was divided into 365 and 1/4°, to correspond with the number of days in a year). If we separate the circular sky from the middle, half of it is over the ground and the other half is under ground. Therefore, at any moment, we can only see half of the Twenty-Eight Mansions, and the other half is invisible.

Astronomers made a celestial globe to illustrate the principle of the diurnal revolution of celestial bodies. For example, Anaximander created a celestial globe with constellation graphics, and a celestial globe named Hun Xiang was designed by astronomers who believed in the theory of sphere-heaven in the Western Han Dynasty. These instruments could vividly demonstrate the rotation of the sky. Figure 2 shows an ancient celestial globe placed in Purple Mountain Observatory. We are supposed to look around from inside this sphere, but it is hard to do so. Now, when we look at the celestial globe, it is just like standing outside the sky to look at the celestial sphere, so the pattern of stars engraved on the sphere is opposite to that we see at night.

Today, we all know that the universe is not the spherical shape perceived by our naked eyes. However, in astronomical research, we often need to consider the sky as a sphere, because only spherical astronomy can make it possible to measure, deduce and explain the motion laws of the sun, the moon and stars. Therefore, it is of great significance for astronomers to realize that the sky is a complete circle scattered with countless stars, and the stars are in their respective positions whether at night or during the day, whether in the sky or underground.

Astronomers use imaginary lines to draw coordinates on the celestial sphere to study astronomy, and these coordinates are almost the same as the longitude and latitude lines drawn on the globe or map we see today. But the difference is that we look at the earth from the outside while look at the celestial sphere from inside. All the large circles passing through the north pole and south pole are called right ascension, and all of them are the same size, with units of hour (h), minute (m) and second (s). The circles perpendicular to the right ascension are called declination, among which the circle equidistant from the south pole and north pole is the largest, which

Fig. 2 Ancient celestial globe

is called the celestial equator. The declination gets smaller and smaller while moving north or south, and it will become two points at the south pole and north pole. All the declination circles are parallel to each other, and all the celestial bodies revolve around the earth every day along with the declination. The units of the declination line are "degree (°)", "minute (')" and "second (″)". With these lines on the celestial sphere, it is convenient to mark the position of each star. Take the famous Vega, also known as α Lyra, as an instance, its position is, right ascension: 18 h 36 m 56.3 s, declination: $+38° 47' 1.0''$. Although the right ascension and declination are imaginary lines, they are drawn on the celestial sphere and rotate with it.

Astronomers also need some imaginary circles to study the motion of celestial bodies. A celestial axis passes under our feet while we, as observers, are standing in the center of the celestial sphere, connecting the northern and southern celestial poles. From where we stand, a line drawn vertically upward intersects the celestial sphere at a point called the Zenith, and a line drawn vertically downward (through the earth) intersects the other half of the celestial sphere at a point called the Nadir. The big circle passing through the zenith, the nadir and the two poles is called the Meridian. Each celestial body will reach the Meridian twice while making a circle, once reaching the highest point and once the lowest point (underground). The earth under our feet extends to all directions and intersects the celestial sphere to form a circle, which is called the horizontal circle. The horizontal circle is perpendicular to the meridian at their intersection and they are equally divided into two parts. The celestial axis, zenith, nadir, meridian and horizontal circle are only related to the observers and do not rotate with the celestial sphere though they are represented on the celestial sphere.

When we visit the planetarium, we often see an instrument composed of many circles, which is called the armillary sphere, and it is the combination of the circles above. Figure 3 is the ancient armillary sphere on display at the Purple Mountain Observatory. Figure 4 is a schematic diagram of the right ascension and declination on the celestial sphere. Figure 5 is a schematic diagram of the meridian and horizon. The northern and southern celestial poles connect the two sets of circle systems.

Fig. 3 Ancient armillary sphere

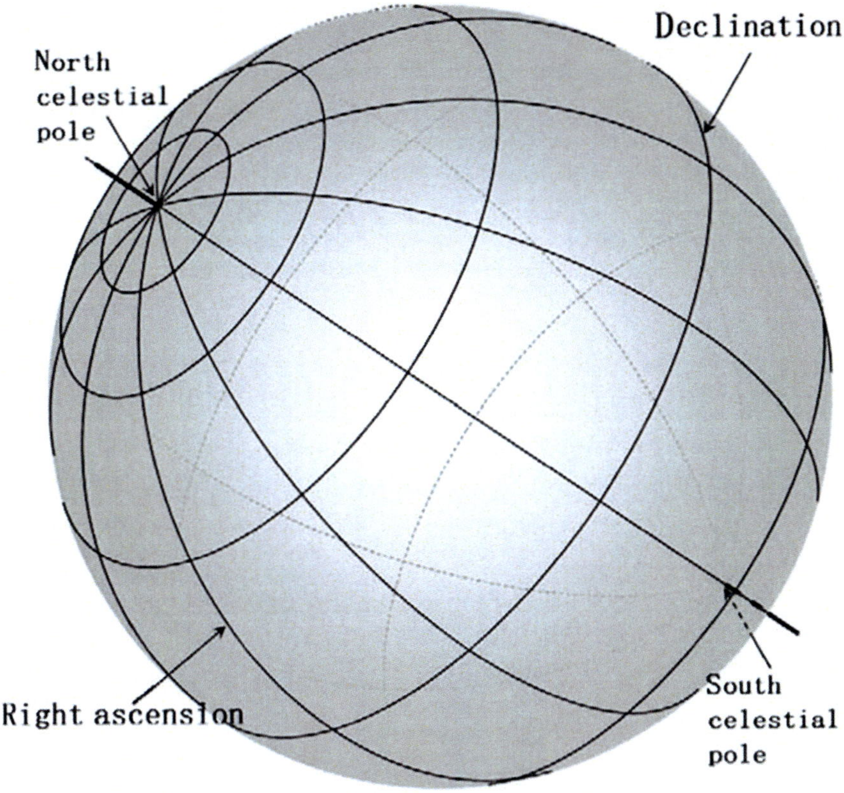

Fig. 4 Right ascension and declination on the celestial sphere

These important circles and coordinates on the celestial sphere were not created by an astronomer on a certain day. They are the results of long-term experience and observation, analysis and research by generations of astronomers. An appropriate understanding of these circles will help readers to better comprehend the knowledge after.

2 Where on Earth Is the Sun?

The sun must be somewhere in the starry sky at night since there are stars in the daytime. The precise position of the sun in the sky will be easily pointed out if we can see the sun and stars at the same time. Unfortunately, it is impossible, because the sun and stars seem irreconcilable and they will never appear at the same time. So, it is not easy to determine the precise position of the sun. Is there any way to do it?

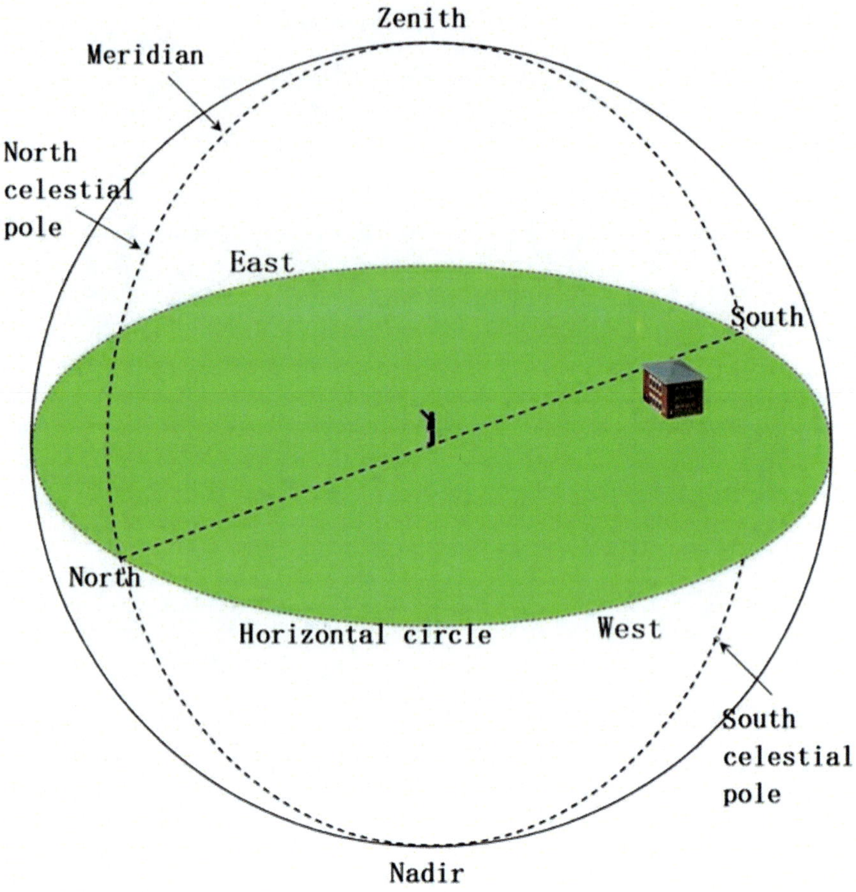

Fig. 5 Horizontal circle and meridian

On a summer evening, when the sun sets and darkness descends, stars appear in the sky. We see Leo (Fig. 6a) near the western horizon. The next morning, we see Gemini (Fig. 6b) near the eastern horizon when it is almost dawn. Based on these, the position of the sun at that time can be roughly determined, that is between Leo and Gemini. What is the constellation between Leo and Gemini? Astronomers have been familiar with the sky after long observation and know that it is Cancer that comes between Leo and Gemini, and the sun is in Cancer during this period.

But the constellations around the sun are not fixed. A month later, when the sun sets, it gets dark and stars appear, people find that Leo has disappeared, and Virgo is near the western horizon. It seems that the sun has moved to Leo. People will see Leo appear on the eastern horizon before dawn in about another month, that is to say, the sun in the west of Leo before has moved to the east of Leo, in Virgo now. By observing the constellations near the western horizon at sunset and the constellations

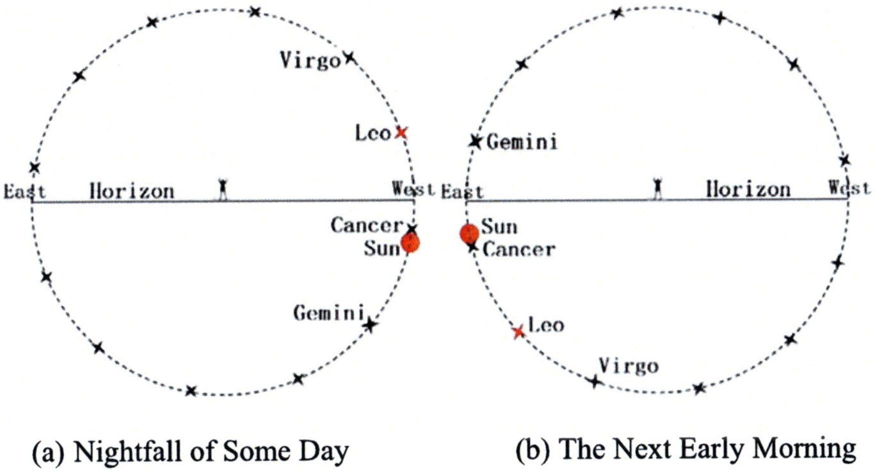

Fig. 6 Determining position of the sun

on the eastern horizon before sunrise the next morning, people are capable of roughly determining the position of the sun, and thus discover that the sun moves slowly from west to east in the background of stars.

The position of the moon in the sky can be observed directly, and the moon is the only visible celestial body in the daytime besides the sun. Astronomers can calculate the position of the sun in the starry sky by observing the position of the moon in the starry sky, especially the position of the moon during the lunar eclipse, because the sun and the moon are 180° apart when the eclipse happens. If the moon is in Taurus, then the sun must be in Scorpio, and the sun must be in right ascension 6 h if the moon is exactly in right ascension 18 h. Of course, astronomers have also come up with a variety of methods (including the heliacal rising method, opposition method, full moon method and culmination method) to infer the position of the sun in the sky at any time as accurately as possible.

By connecting the positions of the sun at all times with imaginary lines, an enclosed circle is formed on the celestial sphere, which is called the ecliptic in astronomy. Generally speaking, the ecliptic is the path of the sun moving slowly from west to east on the celestial sphere.

If the sun does not move in the starry sky, people will never see the constellations that are close to the sun, such as Cancer in the above example. It is because of the movement of the sun that we are conscious of the fact that the stars rise a little earlier than the day before. As time goes on, there will be the cycle of the so-called spring sky, summer sky, autumn sky and winter sky, and thus people have the opportunity to see and understand the stars in the whole celestial sphere.

Please note that "the sun makes a circle in the sky" often mentioned later means that the sun moves along the ecliptic from west to east in the background of the starry sky, which is different from the sun's everyday rise in the east and set in the west. The sun rises from the east and falls in the west, making a circle around the earth in a day, while it takes a year for the sun to move in the celestial sphere from west to east.

As far back as around 2000 BC, the Sumerians living in Mesopotamia had developed the concept of the ecliptic, which ancient Chinese became aware of in the Eastern Han Dynasty at the latest.

The ecliptic obliquely intersects the celestial equator on the celestial sphere, and they are equally divided by each other. While moving on the ecliptic, the sun sometimes reached higher declination and sometimes reached lower declination, which is the reason why the rising and falling points of the sun keep moving and why the sun at noon is sometimes far away from the zenith and sometimes close to it. Figure 7 is a schematic diagram of the ecliptic on the celestial sphere. The ecliptic, like the right ascension and declination mentioned above, is drawn on the celestial sphere and rotates with it.

The twelve constellations of westerners are distributed near the ecliptic. A lot of young men know their star signs. For example, the person born between June, 22nd

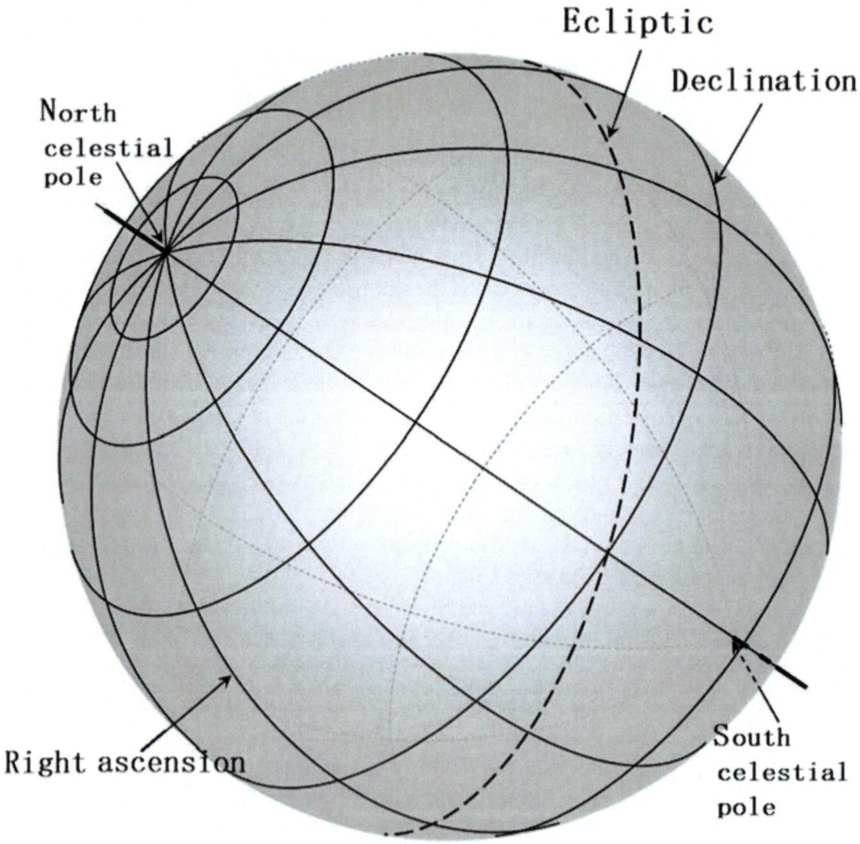

Fig. 7 Ecliptic

and July 22nd is a Cancer. But that doesn't mean that Cancer can necessarily be seen during this period; on the contrary, the sun is in Cancer at this time, so people cannot see Cancer.

It takes a year for the sun to make a circle around the ecliptic on the celestial sphere, and it moves about 1° every day if the large circle is divided into 360°.

As illustrated earlier, the moon and planets also travel from west to east in the background of the celestial sphere, and their paths are near the ecliptic. The seven celestial bodies (Sun, Moon and five planets) are constantly revolving around our earth, some fast and some slow. The fast celestial body will catch up with and surpass the slow one at regular intervals, and it is called conjunction in astronomy when the two celestial bodies reach the same meridian position, such as the conjunction of Venus and moon, the conjunction of Mars and Saturn. If the conjunction of planets and the moon occurs at night, it can be seen with naked eyes by people, but we can only work out the time of the conjunction of planets and the sun as well as the conjunction of the moon and the sun by calculation. It is the beginning of a new (lunar) month in the Chinese lunar calendar when the moon and sun are in conjunction.

The motions of the moon and planets in the background of the celestial sphere can be directly observed by the naked eye, while the position and motion of the sun in the same background are never visible to the naked eye. And there is no visible ecliptic line in the night sky. All of these are completely based on the observation and inference of astronomers. It is a great achievement in the early human history of astronomy to infer the position of the sun on the celestial sphere, the route it moves in the background of the celestial sphere, and the time when the planets and the sun are in conjunction and when the moon and the sun are in conjunction.

3 Variations of the Moon

The shape of the moon changes every day. Sometimes it is like a boat, and sometimes like a golden disk. At first, people thought that the moon is changing its size every day, sometimes thin and sometimes full, and they were lost in admiration for the magic of nature. Did the size of the moon really change? After careful observation, people found that whenever the crescent moon appeared, there was faint light on the other side of it, and they formed a complete circle. This suggested to people that the moon is likely to be a complete circle despite its crescent shape.

When the moon moves slowly from west to east in the sky, it will sometimes block some stars. For example, when the first quarter moon is approaching some star, the star is suddenly blocked before contacting with the moon and the blocked area is exactly the shape of half the moon. So it is not hard to speculate that the part blocking the star is the missing part of the moon. This process proves that the moon is a complete circle all the time though it looks like half of a circle.

Since the moon is round, why does only a part of it shine most of the time? After numerous repeated observations, people found that no matter how the shape of the moon seemed changed, its bright part always pointed to the sun and the dark part always turned its back to the sun. When the moon is full, the whole circle is facing the sun and observers. It seems that the moon itself does not shine, because its shape will not change if it shines. Therefore, it can be inferred that the light of the moon comes from the sun, and the reason why the missing side of it is invisible is that this part is not illuminated by the sun. This should be very logical reasoning.

The Greeks made it clear that the moon shines by reflecting the light of the sun in the 5th century BC. In the 4th century BC, the Chinese realized that the moon shines because of the sun. In the Eastern Han Dynasty, also the beginning of the 2nd century, the famous astronomer Zhang Heng wrote in *Lingxian* that moonlight is a reflection of sunlight.

There is another question. Since the moon is always round, is it a ball or a disk? No matter what shape of the moon we see, the boundary between the shadow part and the bright part is always curved (Fig. 8), which is a phenomenon of a sphere illuminated by parallel light. If the moon is in the shape of a disk, the moon, as can be seen from Fig. 9, will face the earth with a dark side for half of a month, so we cannot see it. While during the other half of the month, the moon will face us with the bright side, so we can see that it is always round in this period. Besides, the moon will suddenly turn invisible or turn bright (the upper position and lower position in Fig. 9), which contradicts the truth. Only the spherical moon will change gradually under the illumination of the sun, from the crescent to the half-moon and to the full moon, then to the half-moon and to the crescent, which shows that the moon can only be a sphere.

We have no idea how long this question puzzled the ancients. One thousand years ago, Shen Kuo, a scientist of the Song Dynasty in China, gave an experimental example to convince others in his *Dream Pool Essays*. Paint half of a black ball with whitewash, and then make it rotate slowly at a certain distance. In the beginning, the black part is facing the observer, then the white part gradually comes into view, which looks like a hook. Growing bigger, the white part becomes a semicircle and eventually a complete circle, and then it slowly shrinks into a semicircle, a hook until the white part is invisible. This process is the same as the changes of the moon seen by people, so we can know that the moon is a sphere.

In ancient times when there was no electric lighting, the moon was very important to human life. Once a lunar eclipse occurred, whether partial or total, it would draw attention. The ancients must be surprised, wondering what covered the moon. Is a sky-dog eating the moon? Is the evil force attacking the moon?

Fig. 8 Curved boundary between the shadow part and the bright part of the moon

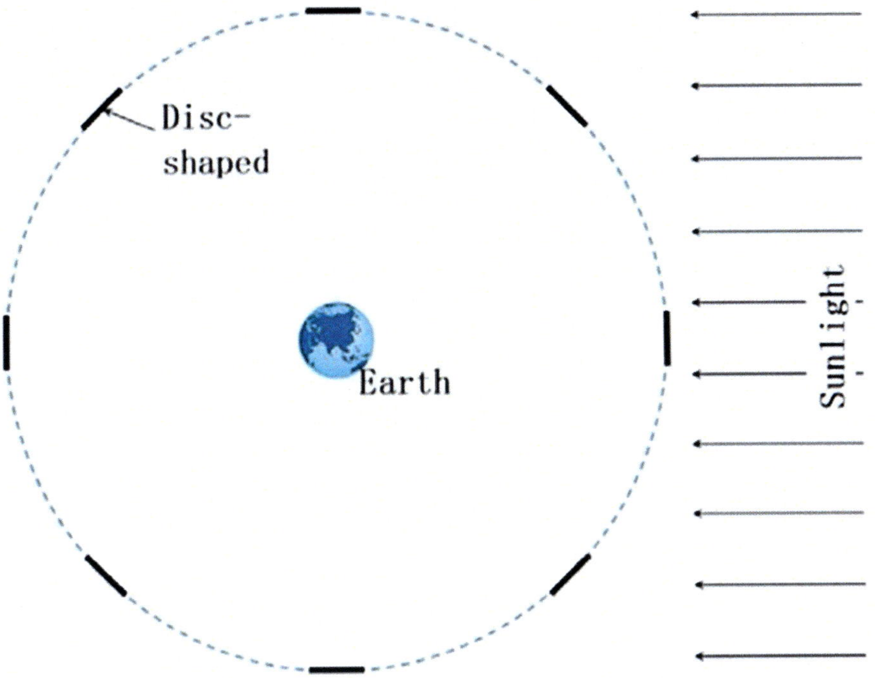

Fig. 9 Disc-shaped moon's changes of brightness and darkness

If a phenomenon occurs many times, its law will be discovered gradually, and it is the same with the lunar eclipse. The moon itself is not able to shine, so there must be something that blocks the light from the sun to the moon. People find that the lunar eclipses always occur at the full moon, when the moon and the sun are in a straight line with observers. And at this time, the moon and the sun are on different sides of us and we are between them. Therefore, some people think that it is the earth under our feet that blocks the light of the sun. When a lunar eclipse happens, the shadow blocking the moon always appears in the east of the moon and fades away at its west, which is consistent with the moon moving from west to east and entering the shadow of the earth.

However, the solar eclipse is more surprising and frightening than the lunar eclipse, especially the total solar eclipse. When it happens, the brilliant sun is gradually swallowed by a black shadow, and it becomes dark all around like night, and stars appear in the sky when the whole sun is swallowed. In ancient times, all nationalities in the world regarded the solar eclipse as the anger of gods and the omen of disaster. What is swallowing the sun? It is found that solar eclipses always occur on the day when the moon is not visible, before the new moon and after the waning crescent. In this period, the moon and the sun are on the same side of observers, and the three are in a straight line. The sun is blocked by a round thing during the solar eclipse. During total and annular solar eclipse, it can be discerned that the size of the thing that blocks the sun is the same as that of the moon. And in the process of blocking the sun, the round black shadow always moves from west to east, which is consistent with the motion direction of the moon in the background of the celestial sphere. These phenomena gradually convince people that the round shadow should be the moon that cannot shine.

As far back as Sargon II (about the 9th century BC), Babylonians knew that the lunar eclipse happens at the full moon. In the Western Zhou Dynasty in the 8th century BC at the latest, Chinese people realized the rule that the lunar eclipse occurs at the full moon and solar eclipse occurs on the first day of the lunar month.

In the 5th century BC, Anaxagoras, the Athenian philosopher, gave a correct explanation of the lunar eclipse. In the late Western Han Dynasty, Liu Xiang (at about the end of the first century BC) said in *Wu Jing Tong Yi* that the reason for the solar eclipse is that the moon blocks the sun while moving.

Zhang Heng, an astronomer in the Eastern Han Dynasty, put forward a basically correct explanation for the cause of the lunar eclipse in *Lingxian*. He said that when the sun and the moon were on opposite sides of us and the three were in a straight line (a phenomenon called "opposition" astronomically), lunar eclipses often occurred because the light from the sun to the moon was blocked by the earth.

The moon is a ball that cannot shine by itself, and its light comes from the sun. The lunar eclipse happens because the moon hides in the shadow of the earth, and the solar eclipse happens because the moon blocks the sun. Astronomers found the reasons why the moon changes in shape and why the solar and lunar eclipses happen. Through the observation of solar and lunar eclipses, astronomers could judge whether the previous prediction of solar and lunar eclipses is accurate, whether the inference of the position of the sun in the celestial sphere is accurate, whether they master the motion laws of the sun and the moon, and whether the established calendar is accurate.

4 How Far Are They from Us?

In ancient times, no one knew how far the moon, sun and stars were from us, and how big they were. But it was not difficult for people to understand that the moon and the sun are definitely far away, because no matter how high they climbed to the top of a mountain, they would not feel the moon and the sun get closer to them, and no matter how far they went east or west, they would never reach the place where the sun rose and set. There was a myth in ancient China. A giant man called Kua Fu was determined to catch up with the sun when it was about to set. However, he could not reach the sun though he ran fast. He was thirsty on the way and drank all the water from the Yellow River and Wei River, and he died of hunger and thirst at last. This myth can reflect that ancient Chinese realized the sun was far from us long ago.

It is common sense that everything looks small in distance and big on the contrary. Seen from the top of a mountain, the people at the foot of the mountain look as small as ants. Seen from the foot of a mountain, the peak seems to rise into the clouds.

Seen from a distance, the mountain is not as high as a tree nearby. The diameter of the sun was calculated by a mathematical method in ancient China, which was 1250 Li (1 Li $\approx$ 0.5 km). It shows that the ancients have realized that the sun is very large though the number was far smaller than the actual size of the sun.

So how far on earth are the sun, the moon and stars from us, and how big are they? This is a very curious matter.

Stars all over the sky are light spots whose distance cannot be told with the naked eye. After long-term observation, ancient human beings found that these countless stars, bright or dim, were always in respective positions and their relative positions were fixed. Therefore, these stars have been called "fixed stars" from ancient times, because these numerous stars are just like fixed on the celestial sphere, revolving around us as a whole. The sun, the moon and the five planets travel slowly in the background of the celestial sphere. Some people speculate that the celestial sphere should be the farthest away from us, which is the boundary of the sky. The sun and the moon cannot move outside the celestial sphere, so they should be closer to us than the fixed stars.

Now, there is a problem to be discussed. Among the sun, the moon and five planets, which is closer to us and which is farther from us?

The reason for the solar eclipse is that the moon blocks the sun, so the moon must be closer than the sun to us. The moon must be smaller than the sun because the moon and the sun look almost the same size in our eyes. And the moon will cover other planets in the course of its movement, which indicates that there is no doubt that the moon is the closest of all celestial bodies to us.

Now the difficulty is how to determine the order of the five planets and the sun. About 4,000 years ago, people measured the time that the sun, the moon and the five planets make a circle in the background of the celestial sphere: 30 years for Saturn, 12 years for Jupiter, 2 years for Mars, 1 year for the sun, Venus and Mercury, and 1 month for the moon. Ancient astronomers intended to get the order of the planets according to the cycles of the celestial bodies. According to astronomers in ancient Greek, the planets far from us have bigger orbits and it takes a longer time for them to orbit the sun, so they seem to move slowly. There was a similar view in ancient China. Zhang Heng, an astronomer of the Han Dynasty, said that the celestial bodies close to the celestial sphere move slowly while those far from the celestial sphere move fast, and the moon is the best evidence.

As a result, Saturn is the farthest, which spends the longest time traveling on the biggest circle. The planet closer to the earth and spending less time traveling than Saturn is Jupiter, and then Mars. The time for the moon to revolve around the earth is the shortest, and it is the closest to the earth and makes the smallest circle, which is consistent with the discussion above. The sun, Venus and Mercury should be between the moon and Mars because it takes one year for all of them to revolve around the earth. Most astronomers agree with this inference, but among the sun, Venus and Mercury, which is nearer the earth and which is farther from the earth? It is a puzzle for the world.

The basic speculation on the distance of the celestial bodies not only laid a certain foundation for the later astronomical research and the establishment of the universe system, but also left a worldwide puzzle for later generations, which is a test on the wisdom of astronomers.

So far, we have explained some important basic inferences made by ancient astronomers through observation and thinking of celestial phenomena. Nowadays, the inferences seem very simple to us. However, in ancient times when productivity was extremely low and science and technology were still in infancy, it is incredibly difficult to realize and comprehend the following inferences: (a) the sky is a complete circle; (b) there is a south celestial pole under the earth, corresponding to the north celestial pole; (c) the moon is a sphere which cannot shine by itself; (d) there are stars in the daytime; (e) to determine the position of the sun and the motion route of it in the starry sky; (f) to understand the cause of the lunar eclipse, and so on. These are undoubtedly great events for astronomy in its early stage, which shows that human understanding of the universe has stepped from the stage of visual recording to the stage of reasoning and judgment, because none of the above conclusions can be obtained by direct observation with the naked eye. These can be called the milestones of early astronomy.

There would probably be no astronomy without these basic inferences, which, certainly, also rely on the developments of other disciplines, especially mathematical theory, as well as the invention and improvement of instruments.

We have no idea which inference was made earlier and how long it took to make these inferences, because many of them were made before the written records came into being. But we can imagine that it must be an extremely long time.

It is necessary to emphasize that, in general, these inferences were understood and accepted by people's common sense at that time. However, what will be told later is far beyond people's common sense. As British astronomer Michael Hoskin said, "The history of cosmology is not the easy story of the rejection of absurd ideas in favour of what (perhaps after a little thought) is seen to be patently true, but the heroic saga of the hard-won rejection of the patently true in favour of the absurd." You will understand this deeply in the following narration.

Chapter 4
Puzzling Earth

1 A Big Obstacle

People will have many questions while looking up at the starry sky. One of them is the easiest but very difficult to answer. How does the sun move east underground after setting in west? This is a question that a farmer, a shepherd, or even a child would ask as long as he is a little curious, and it must be a puzzling question that the ancient people (no matter which nationality or country they belonged to) were confused about. People have the same confusion about the moon and stars. How do the moon and stars pass through underground and move so far from east to west? What does it look like underground? How thick is it? Is there a hole under the ground and how big is it? Where will I be if I keep going straight forward? Will I reach the end of the earth? What is out there? Is the sky connected with the ground?

People have made all sorts of inference about it as well as a lot of fairy tales.

Ancient Babylonians believed that the earth was hemispherical, whose center was a continent surrounded by the ocean, and there was a huge pipe under the earth, which the sun moved through every night and appeared the next morning in the east. The ancient Egyptians believed that the world was like a rectangular box. The ground is the bottom of the box and the sky is the cover, and the four sides of the box are supported by the four mountain peaks rising from four corners of the earth. The ancient Indians imagined that the earth is carried by four elephants that stood on the back of a giant turtle. The Mayans of Central America believed that there are thirteen heavens and nine underworlds in the universe, and the earth is carried by giant crocodiles.

There were three theories about the structure of the universe in ancient China, that is, theory of canopy-heavens, theory of expounding appearance in the night sky, theory of sphere-heavens. According to the first theory, the sky must be above the ground. The sky is semicircular and the earth is arched, and the sun, the moon and stars move horizontally in the sky but cannot move below the earth. Therefore, this theory does not answer how the sun, the moon and stars pass through the ground. The second theory is that the sky is an intangible and infinite space, and the sun,

Y. Ge, *From Geocentric to Heliocentric: How Discoveries Are Made*,
https://doi.org/10.1007/978-981-16-7779-3_4

moon and planets move without restriction. But this theory remained in speculative understanding for a long time, and it does not explain the specific movement of the sun, moon and stars, nor does it mention the shape of the earth. The last theory is that the sky is like an egg, and the ground is like the yolk of the egg, living alone in the sky, so the sky is big and the ground is small. This theory contains some message that the ground is spherical, but it has not clearly stated that the ground is spherical from the beginning to the end, let alone giving relevant arguments. In fact, the theory of sphere-heavens agrees more that the ground is a hemisphere with a flat top and a round bottom, which exactly fills the lower half of the celestial sphere. The diameter of the round ground is approximately equal to that of the celestial sphere, and the center of the ground is in Gaocheng Town, Dengfeng City, Henan Province, China (Yangcheng county in history).

Of course, the so-called Babylonians, ancient Egyptians, ancient Indians and ancient Chinese here are probably just a minority of people. Today we know that some people in ancient times had these thoughts because of the words and works they left. It is not difficult to know that there must be innumerable guesses about those questions in ancient times. Even today, in the face of the rising and setting of the sun every day, a kid cannot help asking questions before learning knowledge about the earth from the classroom or books, and he will do his best to think wildly.

However, on close examination, these myths are still hard to satisfy people's curiosity. If the earth is carried by the divine turtles or elephants, what are they standing on? If the ground is supported by pillars under it, what are the pillars standing on? Rocks and soil are heavier than the sea, so how can the ground float on the sea? Does the sea extend around infinitely? Where are the edges? What prevents the sea from overflowing? And what is holding the seawater? How big should it be if there is a hole underground? How big a hole should be for the huge sun and moon to pass through? How about the stars all over the sky? They are in groups and always keep a fixed relationship of position from sunset to sunrise the next day, and can these, as a whole, also pass through the hole? If the ground is supported by several huge pillars underground that the sun and moon can pass through, what about the stars in groups? Why haven't we seen any stars blocked? If the ground floats on the water, how can the celestial bodies, especially the hot sun pass through water?

It is so hard to answer such questions.

Standing on the solid earth and looking around, we will roughly feel that the ground is flat though there are mountains and rivers on it, just like the ancient Chinese poem goes, "Of Mountain Lu we cannot make out the true face, for we are lost in the heart of the very place." However, if we do not know what the ground is like, we will never be able to explain how the sun, moon and stars get through underground every day, to understand the strange whereabouts of the planets, or to figure out the structure of the whole universe. Therefore, we must first understand what the earth is like to further understand the sky.

This is a huge obstacle for astronomy. If we cannot overcome this big obstacle, there will be no substantial progress in astronomy, and we can only make measurement more and more precise at most. It must be admitted that most of the nationalities in the world failed to overcome this obstacle. Limited by science and technology at that time, the activity range of people was still restricted and they could not leave the ground. It was too difficult to demonstrate what the earth looked like.

It is the outstanding philosophers and astronomers of ancient Greece who led mankind to make it. They have demonstrated the shape of the earth and pushed the study of astronomy to a new stage.

Why is it that the ancient Greeks overcome the obstacle instead of the Chinese or people of other ancient civilizations? Is it by chance? Is there anything worth thinking about?

Astronomers in ancient China paid attention to reality and were diligent in observing and recording in detail, which led to the most systematic and abundant astronomical records of the world, as well as the amazing calendar, astronomical instruments and calculation and prediction of celestial phenomena in ancient China. However, ancient Chinese astronomers rarely asked "why", which may have something to do with the views of some sages in ancient China. For example, Confucius only paid attention to society, never talked about nature, let alone wind and rain, thunder and lightning, the sun, the moon and stars. Xunzi said that a gentleman was still a gentleman even if he did not know the reasons for the evolution of nature, while a villain was still a villain no matter how much he knew. Without knowing these, craftsmen were still skillful in work and officials could continue engaging in politics. The laws would be disorganized if emperors or vassals were interested in these theories. Ordinary people would cause trouble if they were fond of these theories. Ruan Yuan, a famous scholar in the Qin Dynasty, said that the causes of the astronomical phenomena were beyond human ability, so it was only necessary to make clear the celestial phenomena, rather than to seek the reason of why.

Besides, another important feature of astronomy in ancient China is that none of the astronomers were engaged in astronomical research individually, but as members of a special organization under the control of the king. The government even issued laws to prohibit individuals from studying astronomy privately. It is said that there were special astronomical officials named Huozheng in the era of Yao (about 3000 BC). The emperors wanted astronomical officials to make calendars and predict the good or ill luck of the country and the king, and asked for the guidance of action. As for the real cause of astronomical phenomena, they did not care about it, and they did not need the astronomers to seek the real cause of these phenomena.

This may be the important reason why astronomical research in ancient China has always been restricted to practical purposes.

Ancient Egyptians, Mesopotamians and Indians also stopped in face of all kinds of myths and did not continue to seek the answer. Both historical and psychological research show that it is normal that modern science cannot develop from primitive science, and it will be unusual if it does.

2 The Earth Is Under Our Feet

Ancient Greece is located in the east of the Mediterranean Sea, the junction of Europe, Asia and Africa, and there are many bays and islands. In the 5th and 6th century BC, there were highly prosperous economic lives, developed navigation industries and extensive commercial and cultural communication with the surrounding areas. It was the place where the great culture of ancient Egypt and Mesopotamia met. Various views, languages, ideas and myths competed fiercely with each other here, prompting and stimulating people to think and debate.

Ancient Greeks were a nation of navigation and commerce. The characteristic of commercial activities was a fair trade. The status of both sides of the trade was generally equal and it was free to choose who to trade with. The crisis of trading between strangers was prominent, and in the absence of authority, the ancient Greeks made contracts freely as the solution. Therefore, the outstanding aspect of the ancient Greek spirit is to attach importance to equality, contract and individual freedom.

In the ancient Greek city-states, customs differed from island to island and political systems varied greatly. Many different political systems were practiced and developed here, and no centralized rule could unify the customs and cultures of the islands, which opened the door to the free flow of ideas. For example, when different religions confronted each other here, each saying that the gods they believed in were the highest, people had reason to suspect that at least one of the gods they believed in was false. At first, they must deny other's gods, and then some doubted if the gods they believed in were true.

It is under this condition that great thoughts come into being. The universe is no longer the stage of gods. Natural phenomena have their own natural cause, internal order and laws of development and change. People can gradually understand these laws through learning, thinking and reasoning.

The philosophers, scientists and artists in ancient Greek treated others' views and theories with the spirit of questioning, criticism and rational attitude. They openly and boldly put forward their own views and theories, including immature conjectures, and also accepted the questioning, criticisms and judgment of peers and the public at the same time. They pursued strict logic, perfect deductive reasoning system to organize knowledge, and they also pursued the harmony, simplicity and unity of theories. They did not care whether these theories have practical value or not, and they even advocated the uselessness of theories better. It is such environment, thoughts and attitudes that brought about academic innovation and scientific prosperity.

Two anecdotes can vividly depict the style of philosophers and scientists in ancient Greece.

Thales, an ancient Greek thinker, scientist and philosopher who lived around 600 BC and was known as the father of science and philosophy, was also a businessman. However, in the eyes of ordinary people, Thales was not concentrating on business or making money; instead, he was always seeking something impractical, so he could not earn money and get rich. Some people thus said that philosophers were poor people who were useless and could not make money. Once one occasion, Thales used his knowledge to foresee a good harvest of olives in that year, and then he rented all the olive pressing machines in this area. When the olives were harvested, the machines were in short supply, so he took the opportunity to raise the price of the machine and made a lot of money. Thales did it not to be a rich man, but to answer the sarcasm about him from some people. He wanted to tell others that making a fortune is not necessarily harder than studying astronomy, and scientists are capable of making money, but they have more significant things to do.

On another occasion, Thales accidentally fell into the ditch while looking up at the sky. And a maid laughed at him for he just cared about what was happened in the distant sky but could not see what was happening near his feet.

It does not matter whether these two stories are true or not. They are still believable if happen to other scientists, but it is more appropriate to Thales, because ancient Greek philosophers and scientists are exactly such a group of people who look up at the sky and persistently pursue the truth of nature and society.

Pythagoras, who is more than forty years younger than Thales, is a philosopher and mathematician in ancient Greece. Pythagoras visited many places when he was young, and also visited Thales. After his fifties, Pythagoras set up a secret society in southern Italy. There were men and women who have equal status in this society, and all property belonged to the public. The members, who had simple food and strict training, shared common philosophical beliefs and political ideals. Abide by many norms and precepts, they swore never to reveal the secrets and theories of their school.

Pythagoras believed that the number was the origin of all things. The first principle of nature was the principle of "number", and everything could be explained by numbers. All kinds of laws, from music to celestial bodies, could be expressed by mathematical formulas. God ruled the universe by numbers. They believed that their souls could be sublimated and integrated with God by mathematics.

Starting from the aesthetic concept, Pythagoras believed that the universe must be harmonious and simple, and the shapes, as well as orbits of all celestial bodies in the universe, should be perfect. According to him, the sphere is the most beautiful of all the solid figures because any point on the spherical surface is the same far from the center of the sphere. Therefore, the celestial bodies must be spherical, including the sun, the moon, the celestial sphere and the earth. The most beautiful plane figure is the circle. As a result, the celestial bodies should move uniformly in circumference. Pythagoras is the first to deduce that the earth is a sphere.

This deduction seems incredible at first glance, but the truth can be obtained with careful thinking. This view implies a premise that the earth is also a celestial body. What are the shapes of celestial bodies in our eyes? The sun and the moon are round, and the stars are small dots or maybe round though they cannot be seen clearly. On the contrary, it will be more incredible if we think of celestial bodies as other shapes such as triangles and rectangles.

Pythagoras's reasoning is aesthetic rather than scientific. Though many people agree with this point of view, this view alone obviously cannot convince more people. However, scientific reasons were discovered one after another soon. Anaxagoras, an Athenian philosopher in the 5th century BC, pointed out that when the moon, the

Fig. 1. Process of the lunar eclipse

earth and the sun were almost on a straight line, the shadow of the earth would block some of the sunlight to the moon, and a lunar eclipse thus occurred. During a lunar eclipse, the outer contour of the shadow on the moon's surface was always in the shape of an arc, which should be the shadow of the earth. It could be inferred that the earth must be spherical. (Fig. 1)

Aristotle, a famous philosopher over one hundred years after Pythagoras, is an encyclopedic scholar who left his own summary and pioneering ideas in almost all the academic fields of ancient Greece. He argued from the perspective of philosophy and aesthetics, "The shape of the heaven is of necessity spherical... Every plane figure must be either rectilinear or curvilinear. Now the rectilinear is bounded by more than one line, the curvilinear by one only. But since in any kind the one is naturally prior to the many and the simple to the complex, the circle will be the first of plane figures... And the sphere holds the same position among solids. For it alone is embraced by a single surface, while rectilinear solids have several." Ancient Greek philosophers always liked to demonstrate from philosophy though it was abstract for ordinary people.

Of course, experimental examples are also essential. Aristotle also cited the example of the lunar eclipse, "The evidence of the senses further corroborates this. How else would eclipses of the moon show segments shaped as we see them." "Since it is the interposition of the earth that makes the eclipse, the form of this line will be caused by the form of the earth's surface."

No matter where the moon is (in the first or second half of the night) when the lunar eclipse occurs, it will always look like a part of the circle when the shadow of the earth sweeps across the moon. This indicates that the shape of the earth is such that it can cast a circular shadow in any possible range, so the earth can be only spherical.

Astronomers in ancient China also made correct explanations for solar and lunar eclipses. Lunar eclipses invariably occur at the full moon, when the moon and the sun are in a straight line with us observers. We are between the sun and the moon and it is the shadow of the earth that blocks the moon. It is a pity that they did not continue to analyze the shape of the shadow, and they thus did not conclude that the earth is spherical.

Aristotle also gave another example. We see different stars while moving north or south. It is true that some stars can be seen near Egypt, but not in Greece or further north. Moreover, some stars that are completely visible in the north and always above the horizon (the circle of perpetual apparition) will fall into the horizon in the south. Therefore, the shape of the earth must be round, and the volume of this sphere is not very large, otherwise there will be no such obvious difference in the case of a small change of position. As can be seen in Fig. 2, the ranges of stars seen at point A and point B are different, and the height of the north star is also different. For example, the north star seen at point A is much higher than that seen at point B.

In the 12th year of the Kaiyuan Region Period of the Tang Dynasty, Nangong Yue and Monk Yixing led a large-scale measurement of the difference between the shadows in the north and in the south, which completely rejected the hypothesis that the shadow of an eight-foot-high pole at noon of the summer solstice is one inch shorter when it moves 500 km due south and one inch longer when it moves 500 km due north. Meanwhile, they had two findings. The first was that from the south to the north, the height of the North Pole gradually increased, and the height difference of the North Pole was in a fixed proportional relationship with the distance moved between the north and the south, that was, the height difference of the North Pole was 1° when the distance moved between the north and the south was 351.27 miles. The second was that looking south from Jiaozhou in China in ancient times, they could see many stars that were invisible in the Central Plains, and these stars were all in

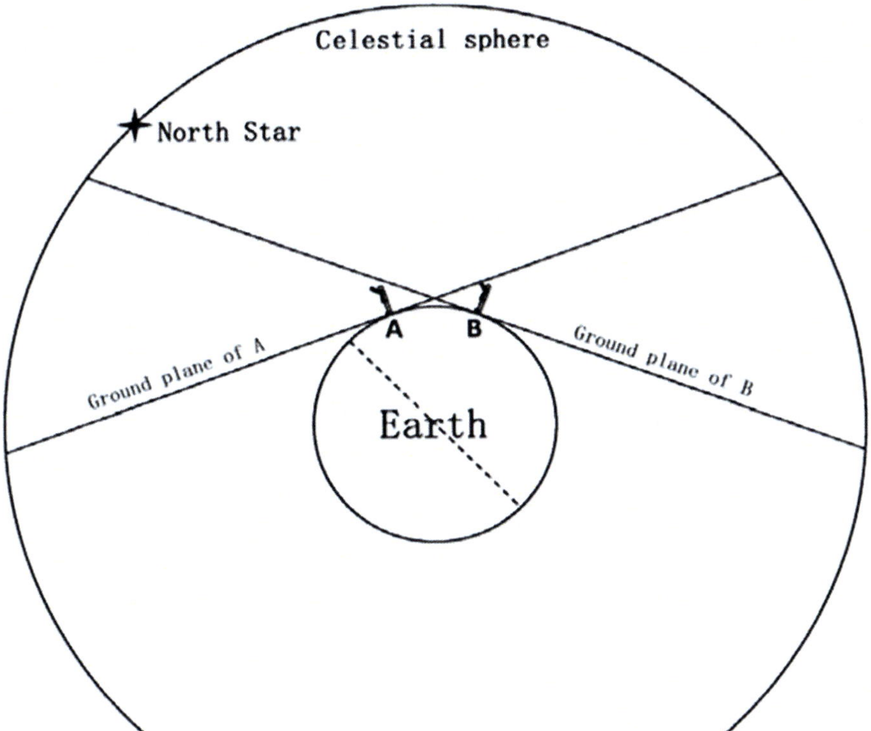

Fig. 2. Height changes of the north star

the circle of perpetual occultation drawn before by astronomers who believed in the theory of sphere-heavens. The two facts show that the earth should be arc-shaped at least in the north-south direction, but it is a pity that Monk Yixing and his fellows did not think of this. In the Yuan Dynasty, Guo Shoujing, along with others, organized a larger survey of the height of the North Pole. A total of fourteen positions from the South Sea (15° high in the North Pole) to the North Sea (65° high in the North Pole) were selected for measurement, but they also failed to understand that the earth is circular. In their view, the height of the North Pole was just what it was, not anything else, and there was no need to think more or ask "why". So Chinese astronomers once again missed the opportunity to discover by themselves that the earth is spherical.

The ancient Greeks provided a third argument for the fact that the earth is spherical. That is, why do people first lose sight of the hull of the ships sailing far away, and then the sails of them?

If the earth is flat, a sailing boat will get smaller and smaller in people's eyes as it goes farther and farther away from the coast, and it will become a little dot before disappearing at last. However, this is not the truth. In the beginning, those who watched the sailing boat set off could see the whole boat, the lower hull as well as the upper sail. But after some time, the hull gradually disappeared and seemed submerged by the sea, leaving only the sails, then the top of the sails, and finally the whole sailing boat disappeared (Fig. 3).

No matter which direction a sailboat sails, the north, the south, the west, the east, or any direction, the bottom of it disappears first. Besides, sailboats always disappear at about the same speed. Part of the hull will be blocked when it goes two miles away, no matter what direction it goes. It seems that in any direction, the earth is curved and the curvature is equal, so it can be inferred that the earth is spherical.

Perhaps the acquisition of this evidence is closely related to the geographical environment of ancient Greece. Ancient Greece was located in the east of the Mediterranean Sea, with many bays and islands as well as a developed navigation industry. The sea and ships were the most common, so it was easy for people to see that ships gradually disappear from bottom to top while watching their families sail away and looking into the distance in hope of their return. China and the other three cradles of

Fig. 3. Gradual disappearance of a sailing boat

civilization were all located in the alluvial plain and people mainly lived inland, so it was difficult for them to see the ship going away. Certainly, it doesn't mean that we can realize that the earth is a sphere as long as we see the disappearance of ships. The gap here is huge and even insurmountable for most people.

Moreover, there is another fact that proves the earth is spherical. The sun rises at different times in two places far apart, east and west, and the sun rises earlier in the east than in the west. If the earth is basically flat, people in the east and west will see the sunrise at the same time. How did people find that the sun rises at different times in two places far apart? It was not easy to discover in ancient times when there were no clocks. Fortunately, the lunar eclipse provided such an opportunity because it happened at the same time for all people who could see the moon on the earth. People in the east and west saw a lunar eclipse happen at different times. For example, when people in a certain place saw a lunar eclipse at midnight local time, residents living hundreds of kilometers east saw it after midnight local time, and residents living hundreds of kilometers west of them saw it before midnight local time. When people from the east and west got together to talk about the same lunar eclipse, they would find their "local time" were different. So, it was found that the "local time" of the two places with long distance in the east-west direction was different, which meant that the sun rose at different times in the two places. Why? Because the earth is spherical, and when the sun's position changes, there will be different local times to see the sun in the east and the west. People in the east meet the sun earlier and see the sun disappear earlier than people in the west.

Therefore, the earth is spherical, hanging alone in the universe instead of being carried or supported. The sun, the moon and stars do not pass through "underground", but revolve in their respective circular orbits.

Undoubtedly, this is the first important inference in astronomical history that goes against human's common sense. If the earth under our feet turns out to be spherical, then, aren't the people on the other side of the earth head down? Won't they fall down? It is really incredible. Besides, how can the earth hang alone in the universe? Won't something without support fall down? People at that time must be very puzzled about these. In this regard, Aristotle argued that matter has the nature of tending toward the center. Upward means to leave the center, while downward means to reach the center. "Heavy" means something that moves naturally towards the center, and "light" means something that moves away from the center. Perhaps this was the best theory and explanation at that time.

Of course, there would be other doubts and puzzles. But generally speaking, the argument proving the earth is spherical was reasonable, and the spherical earth was also the soundest explanation for the question that how the sun, moon and stars pass through the earth. So, around the 3rd century BC, few thinkers doubted the inference that the earth was round, and the theory of round earth became a final conclusion in ancient Greece and the areas affected by its civilization. As for those doubts, thinkers believed that they just hadn't found a reasonable explanation for the time being.

Almost at the same time, the ancient Greeks realized that the earth was very small compared with the celestial sphere. It is not easy to understand this, but it is also easy, because the following proof can be comprehended with a little bit of geometric

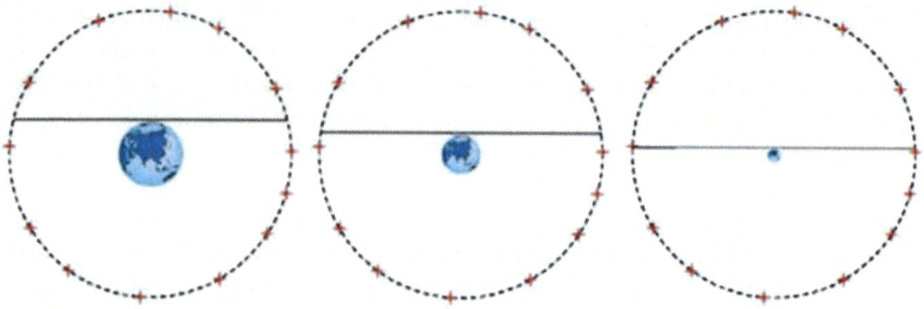

Fig. 4. The small earth compared with the celestial sphere

knowledge. What we can see at any time and any place is exactly half of the celestial sphere, that is to say, the horizontal circle made anywhere on the earth always bisects the whole celestial sphere (and the celestial equator). If the scale ratio of the earth to the sky is not very small, the plane made through any points on the earth's surface will only make the lower part of the celestial sphere larger than the upper part. Only when the earth is almost a point can we see exactly half of the celestial sphere anywhere, as seen clearly in Fig. 4. Therefore, the earth must be very small compared with the celestial sphere, and Aristotle also believed that the earth must be at the center of the universe accordingly.

3 How Big Is This Sphere?

If the earth is flat, some people may wonder how big the plane is and how far it can extend. Now, some also wonder how big the sphere is after we all know that the earth is spherical. The answers like that the earth is very large or that it is infinite could not satisfy human beings. But in the era when there were no modern advanced technologies of aerospace, aviation and navigation, or telescopes and sextants, could we calculate the radius of the earth? Yes! We could do it if we knew a little about geometry.

In the 3rd century BC, Eratosthenes, the director of the Alexandria Library in ancient Egypt, wanted to know how big the earth was. He learned that in the southern frontier near the biggest waterfall of the Nile (Syene, Aswan today), the shadow of the cathedral column became shorter and shorter near noon on the summer solstice (the longest day of the year), and finally disappeared at noon. The sun shone directly from above people's heads to the bottom of a deep well at noon. The long upright pole at noon did not cast a shadow on the ground. All these meant that the sun's light was perpendicular to the ground at noon on the summer solstice in Syene. Alexandria, the hometown of Eratosthenes, and Syene were located in the north and south positions of two very close meridians. Eratosthenes made the same observation, but the sunlight did not reach the bottom of the well directly on the summer solstice,

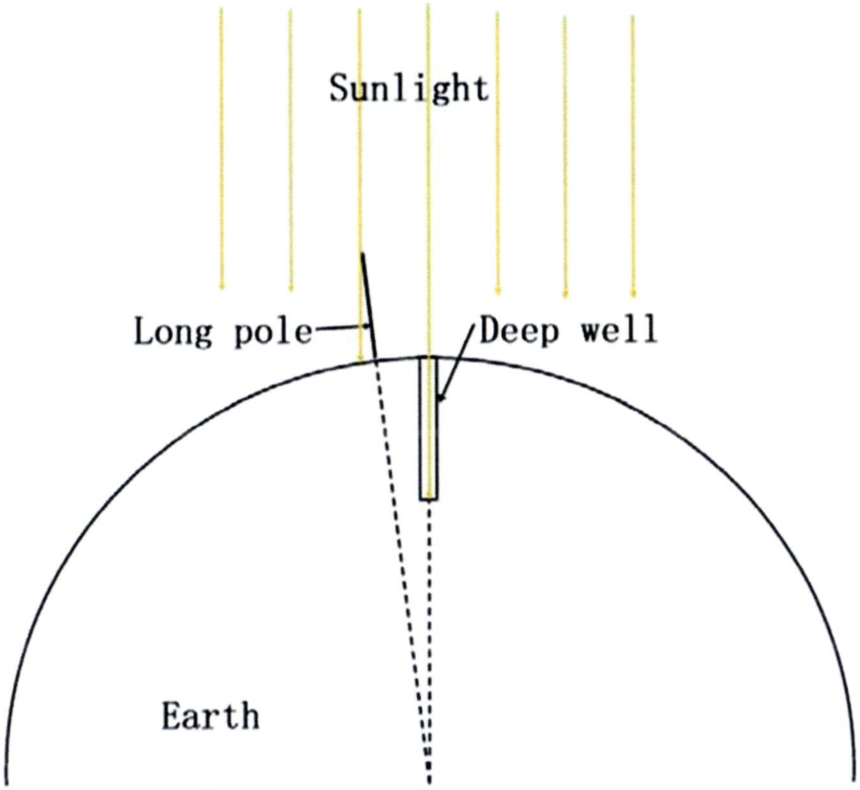

Fig. 5. Measuring the size of the earth

and had a certain angle of declination. The long upright pole cast a short shadow on the ground at noon on the summer solstice, and Eratosthenes measured that the angle between the sun and the pole was 7°. Figure 5 shows Eratosthenes's method of measuring the size of the earth.

These phenomena are not unique to the vicinity of ancient Egypt, but can be seen in many places of the earth. For example, in China, there are almost no shadows of buildings at noon on the summer solstice in Gejiu City, Yunnan Province. While in Ya'an City, Sichuan Province, which is at the due north of Gejiu, if a pole is vertically erected on the ground, the angle between the sun and the straight pole at noon at the summer solstice is also close to 7°. Another example is Chiayi City, Taiwan Province, where there is almost no shadow of buildings at noon on the summer solstice. In Hangzhou City, Zhejiang Province, the due north of Chiayi, if a pole is vertically erected on the ground, the angle between the sun and the pole at noon on the summer solstice is also close to 7°. The problem is that if you do not seize these phenomena and think about them, what can they tell us?

The sun is so far away from us (Aristarchus, an astronomer before Eratosthenes, proved this, which will be described in detail in Chap. 5), so the sunlight can be regarded as parallel when it reaches the earth. Eratosthenes believes that it is because the earth is spherical that the angles between the pole and the sun's light are different in two places, and if the pole is inserted into the earth's core, it will intersect at an angle of 7° with the earth. Seven degree is about 1/50 of the 360° circle of the whole earth. Eratosthenes knew that the distance between Alexandria and Syene was about 5000 stade (unit of length in ancient Greek), and 5000 stade multiplied by 50 equaled 25000 stade, which was exactly the length of the circumference of the earth.

It is said that this length is quite close to the accurate data we know today. But in fact, the data he adopted is relatively rough and it is somewhat coincidental to get such a result, but his ideas and methods are correct. Carl Sagan, an American popular science writer, said in his book *Cosmos*, " Eratosthenes' only tools were sticks, eyes, feet and brains, plus a taste for experiment. With them, he deduced the circumference of the Earth with an error of only a few percent, a remarkable achievement for 2,200 years ago. He was the first person to accurately measure the size of a planet." Therefore, in the era when there were no modern advanced technologies of aerospace, aviation and navigation, or telescopes and sextants, we really could calculate the radius of the earth by using geometric knowledge, and what was more important than geometry is whether we could think of this.

Of course, it was the round-the-world voyage of Magellan's fleet two thousand years later that directly proved that the earth was spherical finally. Magellan's fleet sailed westward in September 1519, and returned to Spain from the east in September 1522 after experiencing all sorts of hardships, proving with facts that the earth is spherical and nothing has been found to support the earth.

Today, we all know that the earth is spherical. While looking at the above proofs, do you think it is simple or difficult? Many people were aware of the shadow during the lunar eclipse, the change of the height of the north star, and the disappearance of sailing ships, but how many people could associate them with the shape of the earth? When there were no special instruments and people never left the ground, why could some people fly to such a height with imaginary wings to overlook the earth, and dare to assert that the earth is a sphere which hangs alone in space?

Compared with the earth, human beings are extremely small, and the range of human activities in ancient times was very limited. It was a great breakthrough in early astronomy to realize that the earth under our feet is a sphere and that this sphere is very small compared with the celestial sphere. Without this breakthrough, the conjecture about the sky and the earth will always be like feeling the way in the dark, so this is an important milestone in the history of astronomy. Only on this basis can the exploration of the universe structure really begin.

Without these outstanding proofs made by ancient Greek philosophers and astronomers, when would humans know that the earth is spherical? Until the end of the Magellan's voyage around the world? But without the previous understanding that the earth is spherical, did Columbus and Magellan dare to sail west? Were they willing to never return?

Chapter 5
Guess What the Universe Is Like

1 Who Can Guess What the Universe Is Like?

All the nationalities in the world, no matter when their civilization started or how fast they evolved, have observed and thought about the sky and the earth, and had all kinds of guesses about the structure of the universe. Certainly, we see that most of the conjectures are ultimately attributed to gods, because all these are so mysterious that our hearts will tremble at the thought of the vastness and permanence of the universe.

When ancient Greeks knew, strictly speaking, demonstrated and believed that the earth was spherical, the study of the model of the universe really began. What is the truth of the universe? Which of the astronomers' conjectures and models are consistent with the truth? The basic components of the model are the earth, the sun, the moon, the five planets and the celestial sphere, and astronomers' task is to "fiddle" with these components, and build their models to see if they can reasonably explain the various astronomical phenomena. In particular, they need to make reasonable explanations for the strange tracks of planets, such as moving forward, moving backward and stationary in the background of the celestial sphere, to see if they can predict the positions of celestial bodies at any time with the help of models. It is believed that only the model with the most reasonable explanation and the most accurate prediction can be the truth of the universe.

As far as people's intuition is concerned, the simplest and most reasonable explanation is that the earth is at the center of the universe, and the celestial sphere, the sun, the moon and planets all revolve around the earth from east to west. They rise in the east and fall in the west, making a circle every day. It must be the first model of the universe in the world if it can also be called a model, a self-evident model of the universe.

However, things are not so simple. With a little careful observation and inference, it is not difficult to find that the sun, the moon and the planets are not completely synchronized with the celestial sphere while following it to rise and fall, and they all revolve slightly more slowly than the celestial sphere. For instance, the sun is about

Y. Ge, *From Geocentric to Heliocentric: How Discoveries Are Made*,
https://doi.org/10.1007/978-981-16-7779-3_5

1° slower every day, and it falls behind a circle a year. The moon is about 13° slower, and it falls behind a circle a month. Ancient Chinese astronomers described it as that the sun and the moon travel slowly and the celestial sphere moves fast. So, in our eyes, the earth, the sun, the moon and the planets move backward slowly from west to east in the background of the celestial sphere which is so weird that they move both west and east, requiring a reasonable explanation.

Both Vitruvius, a Roman architect in the first century BC, and Wang Chong in the Han Dynasty in China in the first century AD, once used an ant as an example to describe the movement of celestial bodies. Figure 1 is a schematic diagram of this analogy. A disk is rotating clockwise and makes one circle a day, and an ant is crawling slowly counterclockwise from point A to point B at the edge of the disk. Suppose the ant makes one circle a month, and the disk rotates while carrying the ant, but the ant crawls slowly at the edge of the disk in the opposite direction. The disk rotates fast, but the ant crawls slowly. In Fig. 1a, the ant is close to point A. After 12 h, as shown in Fig. 1b, the disk has rotated 180°, and the ant has reached a certain position between point A and point B. For the observer standing still in the middle of the disk, the disk and the ant on it rotate clockwise every day, but if you look carefully, you can find that the ant crawls counterclockwise relative to the disk. In this way, the disk moves around the center for a circle a day, while it takes a little longer more than a day for the ant to orbit the center.

This vivid analogy can well explain the apparent motion of the sun, the moon and planets. Attached to the sky, the sun, moon and planets are forced to rise in the east and set in the west with the celestial sphere every day, just like the ant forced to revolve around the center of the disk, and at the same time, they are respectively moving slowly from west to east. It is a puzzle that these ants do not crawl uniformly from west to east, but sometimes faster and sometimes more slowly. Their routes are roughly similar, which are inclined to the circular arc drawn by the rotating celestial

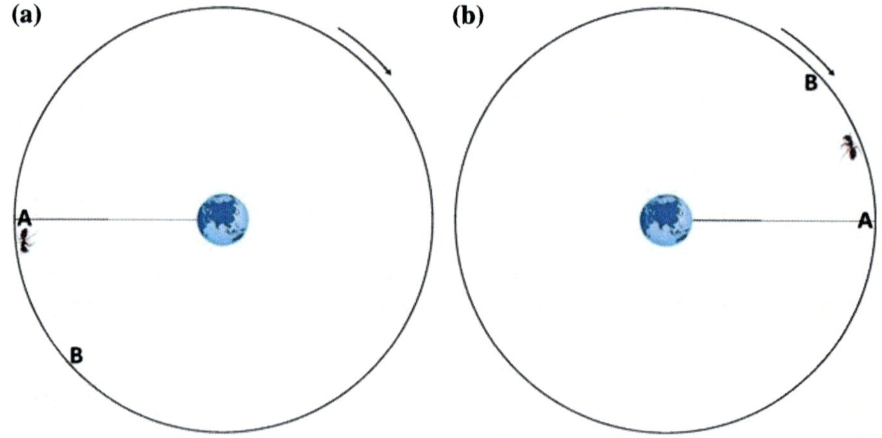

Fig. 1 Ant and disk

sphere. What is even more incredible is the crawling of the five planets, which, besides moving at various speeds as the sun and the moon, sometimes stops, even moves backward for a short time, then stops again, and turns back to move on.

Human beings should really thank the planets, because they will not encounter so many puzzles without the planets. Maybe humans would never be capable of exploring the truth of the solar system if there were no "allure" and "hint" from the planets.

The movements of the five planets are so strange that it is difficult to understand, but we can know that they follow certain laws after long observation. Mercury and Venus are always wandering on both sides of the sun, moving from the greatest eastern elongation (evening star) to the greatest western elongation (morning star), and then from the greatest western elongation to the greatest eastern elongation repeatedly. Both the two planets appear retrograde in the process of moving from the greatest eastern elongation to the greatest western elongation. Saturn, Jupiter and Mars can leave the sun's control and circle the earth. They can revolve to the same position as the sun, which is called the conjunction in astronomy. Moreover, they can also revolve to the position opposite to the sun and face the sun across the earth, which is called the opposition in astronomy. From the conjunction to the opposition, then from the opposition to the conjunction, over and over again, the planets are all moving backward around the time of the opposition. Mercury and Venus are the brightest when they are around the greatest eastern elongation and the greatest western elongation. Saturn, Jupiter and Mars are the brightest at the time of the opposition. The synodic period of Mercury and the earth is about 116 days, Venus and the earth about 584 days, Mars and the earth about 780 days, Jupiter and the earth about 399 days, and Saturn and the earth about 378 days. It takes about 29.5 years for Saturn to make a circle in the background of the celestial sphere, about 11.86 years for Jupiter and about 1.88 years for Mars respectively. Venus and Mercury spend one year circling the earth like the sun because they are always wandering on both sides of it. All these show that the motion of planets follows certain rules, just as the ancient Chinese philosopher said, "Nature is the true law". The movement of the celestial sphere has its own rules which do not exist because of a wise king or disappear because of a fatuous king.

Astronomers have spared no efforts to make many conjectures about the cosmic model that can give reasonable and convincing explanations of the movement of the sun, the moon and the planets, as well as the reasons behind these laws. The following are three major cosmic models that are relatively complete and influential in history.

2 A Fire in the Center of the Universe

In the fifth century BC, Philolaus, an excellent representative of the Pythagorean School in ancient Greece, put forward the central fire universe model. He said the center of the universe is a fire, and the earth makes a circle around the central fire along the west-to-east orbit every day. Outside the earth is the moon, which makes a circle around the central fire along the west-to-east orbit every month. Outside the moon is the sun, which makes a circle around the central fire along the west-to-east orbit every year. Outside these, Mercury, Venus, Mars, Jupiter, Saturn are arranged in turn, and the outermost sphere of fixed stars is still. Figure 2 is the central fire model of Philolaus.

A feature of this model is a counter-earth between the central fire and the earth, which moves around the central fire at the same angular velocity as the earth. The earth always faces the central fire with the same side, while the Greeks live on the side opposite to the central fire. Therefore, people on the earth can never see the central fire or the counter-earth.

What are his reasons to make such a model? People of the Pythagorean School advocates the harmony of the universe. According to Philolaus, the sun, the moon and the planets revolve around the earth from east to west every day with the celestial

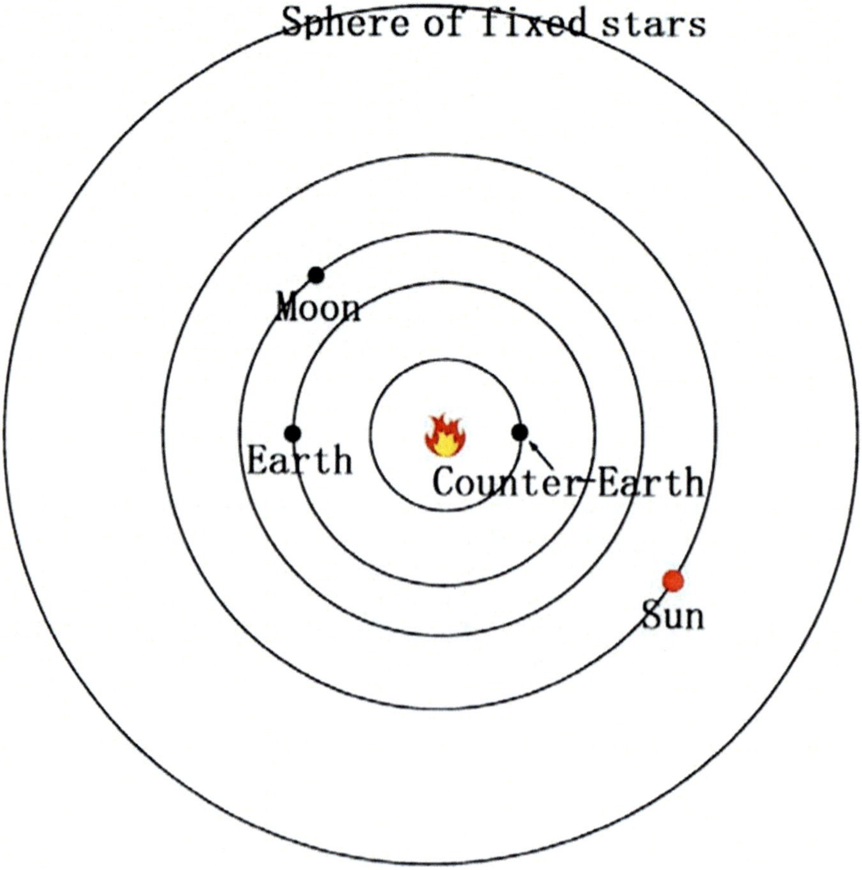

Fig. 2 Central-fire model of Philolaus

sphere, but they also move in the opposite direction, which is disharmonious. How to solve this problem? The only choice is to let the earth make a circle around the central fire every day.

Philolaus, as a member of the Pythagorean School, also firmly believes that number is the origin of everything in the universe, and all phenomena and laws in nature are determined by number, which must obey the harmony of number, that is, obey the relationship of number. Sphere is the most perfect geometry, so the earth must be spherical. The sun, the moon and planets must move uniformly on the circumference because the circle is the most perfect plane figure. He also thought that 10 is the most perfect number, so there must be 10 celestial bodies.

10 is indeed perfect. Everyone has 10 fingers and 10 toes, and $1 + 2 + 3 + 4 = 10$. But there are only 9 visible celestial bodies (including the celestial sphere). What to do with it? As a result, Philolaus created a celestial body called counter-earth to make up 10 celestial bodies. It is funny that he did not consider the realistic observation at all but constructed the universe model completely according to his own belief.

We have no idea what other specific thoughts Philolaus had at that time and why he assumed that there was a fire in the center of the universe, due to the limited information after so long. But we may be capable of inferring some of his thoughts when the two viewpoints above are put together. The earth must rotate to achieve harmony of the universe, but if the earth is placed in the center of the universe to rotate, counter-earth, the tenth celestial body, can be placed nowhere, so the earth must be moved away from the center. Let the earth revolve around the center once a day with its one side always facing the center, then the purpose of making the earth rotate can be achieved and there is also room for the counter-earth. However, it is obviously unreasonable that the earth, the sun, the moon and the planets revolve around a center with nothing in it, so he set a fire in the center, which may be inspired by the brilliant sun. In order to solve the contradiction that there are two fires in the universe, Philolaus explained that the sun is a big mirror and it reflects the light from the central fire.

Today, we may think this hypothesis too simple, even a little ridiculous. But it was extremely valuable to put forward such a universe model over 2000 years ago when science and technology just started and the whole world was dominated by gods.

The main highlight of this universe model is that the earth moves instead of the celestial sphere. The earth, with its one side always facing the central fire, moves around the central fire and makes a circle along a west-to-east orbit every day near the central fire. The effect is similar to that of the earth's rotation, which provides a relatively reasonable explanation for the apparent motion of the sun, the moon and stars. Thanks to this kind of motion, the contradiction that the sun, moon and planets we see revolve around the earth from east to west with the celestial sphere every day and also revolve from west to east against the celestial sphere have been resolved. The celestial sphere and its center keep still, and the five planets, the sun, the moon and the earth revolve from west to east according to the same law. The whole universe appears harmonious.

However, the problem is also obvious. When the earth moves around the center every day, it will be sometimes close to and sometimes far away from the stars, and the apparent position between stars should change, unless the distance between the earth and stars is infinite. But the distance between stars and the earth is finite according to Pythagorean School, so why has it never been observed that the apparent positions of stars change in one day? This is how Aristotle opposed the central fire model 100 years later.

The so-called counter-earth is completely imaginary without any practical basis to make up ten celestial bodies. And where is the basis of central fire? These are all unconvincing.

Hicetas and Ecphantus, scholars of the Pythagorean School, put forward the theory of the earth's rotation based on the Philolaus' model, believing that the earth is in the center of the universe and rotates once a day. Later, Heraclides of Platonism inherited their views, and explained the apparent motion of celestial bodies by the rotation of the earth around its axis. At the same time, he noticed that Mercury and Venus had never been away from the sun, and then put forward that the two planets moved around the sun and then moved around the earth together with the sun. In fact, the distinction between these scholars and Philolaus is that they are not superstitious in the number of 10, so there is no need to consider where the counter-earth should be placed, and naturally the central fire is not needed.

According to Philolaus's successors, the universe model should be adjusted as shown in Fig. 3. The earth is at the center of the universe and rotates around its axis once a day (a sidereal day). What is near the earth is the moon, which revolves around the earth once a month. Outside the moon is the sun. Mercury and Venus revolve around the sun and follow the sun to make a circle around the earth once a year. Outside the sun are Mars, Jupiter and Saturn in turn, and then the stationary sphere of fixed stars.

The modified Philolaus's model is a relatively complete geocentric model, but this model is too simple to explain the strange movements of planets. Why does the brightness of the planets change? Why do they move sometimes fast but sometimes slowly? Why do they move backward sometimes? All of these questions need explanations. What is surely even more unbelievable is how the earth can move. Let alone revolving around the central fire, even rotation is impossible! At that time, people could not imagine that the earth under their feet was moving with their common sense and intuition, so Phililaus's universe model was not widely recognized, but their unique understandings provided a direction for the birth of heliocentric theory in the future.

3 The Earth in the Center of the Universe

A new method should be adopted when one fails. Design another one if one model doesn't work!

More than 2000 years ago, most astronomers believed that the center of the universe must be the earth. Can our intuition be wrong? Only the center can be stationary in a moving universe, so the earth must be in the center of the universe.

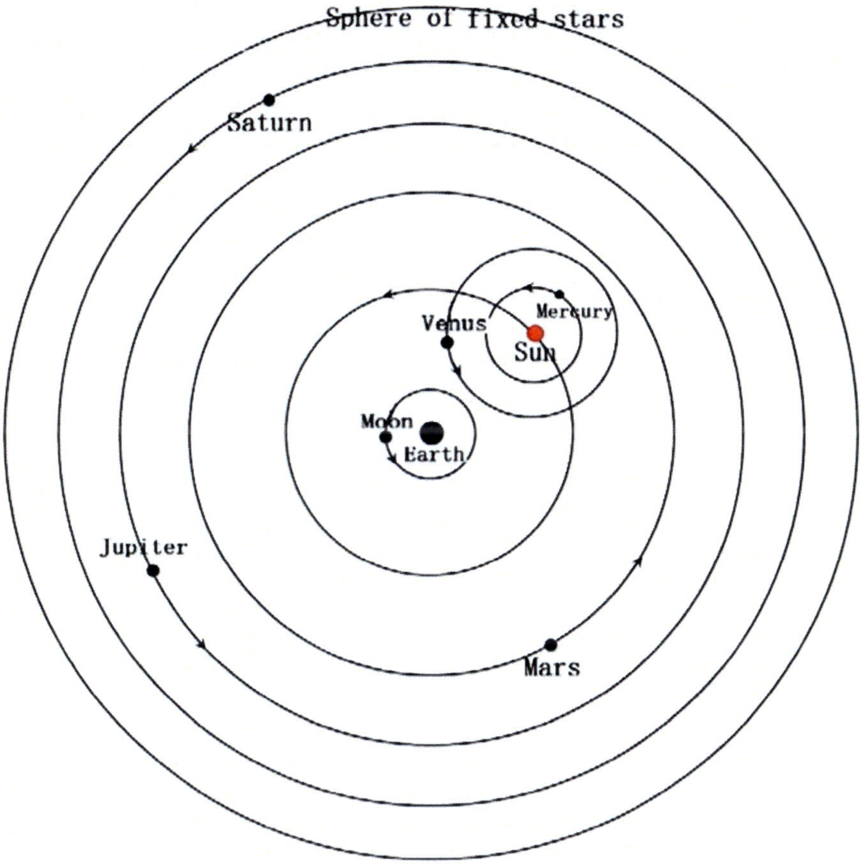

Fig. 3 The modified Philolaus's model

According to Pythagorean School, the universe is harmonious and perfect, and the most beautiful plane figure is the circle, so the celestial bodies must move on the circumference at a uniform speed. This view is correct if both the circle and the ellipse are collectively called the circle, which almost no one doubts, because no other form of motion can be better understood and accepted. But in fact, we see the planets move sometimes fast and sometimes slowly, sometimes in the antegrade direction and sometimes in the retrograde direction. How to understand this? Plato, a philosopher in ancient Greece around the fourth century BC, believed that this is just a superficial phenomenon, which can be explained by the combination of uniform circular motion.

Plato's words pointed out the direction for studying the motion of celestial bodies, especially the complex motion of planets. The movements of the sun and the moon seem to be uniform circular motion (Only long-term observation and comparison can reveal they do not move uniformly, but it is hard to detect with naked eyes). The planet moves secretly, but it will repeatedly return to a certain place at the same interval. It is impossible for a celestial body that moves randomly to be so regular, and only by moving in the circumference will it repeatedly return to a certain place regularly, which indicated that the planet must move on the circumference. Astronomers and philosophers in ancient times, before fully understanding the nature of ellipses, could not imagine what geometric figures that celestial bodies could move on besides circular orbits. They also could not imagine how a celestial body in a circular could move at different speeds.

The task now is to find a combination to make the planet move at a constant speed in a circle though what we see is different. Astronomers tried their best to find the answer.

Plato pointed out the direction, but he did not come up with any good idea to combine. In his book *Timaeus*, he proposed a concentric spherical shell structure model with the earth in the center. The spherical shells of each celestial body are arranged from near the earth to far from the earth in the following order, the moon, the sun, Mercury, Venus, Mars, Jupiter, Saturn and stars. Of course, such a simple description can only be regarded as a draft of the model.

Eudoxus, a student of Plato, developed his teacher's viewpoints and refined his model. He found a combination of circular motion, which could be a good solution to the problem of planets' retrograde motion.

Figure 4 shows Eudoxus's solar motion model. He used the combination of two concentric spheres to explain the annual apparent motion of the sun. The outer spherical shell is the celestial sphere, and the two ends of its axis are the two poles of the

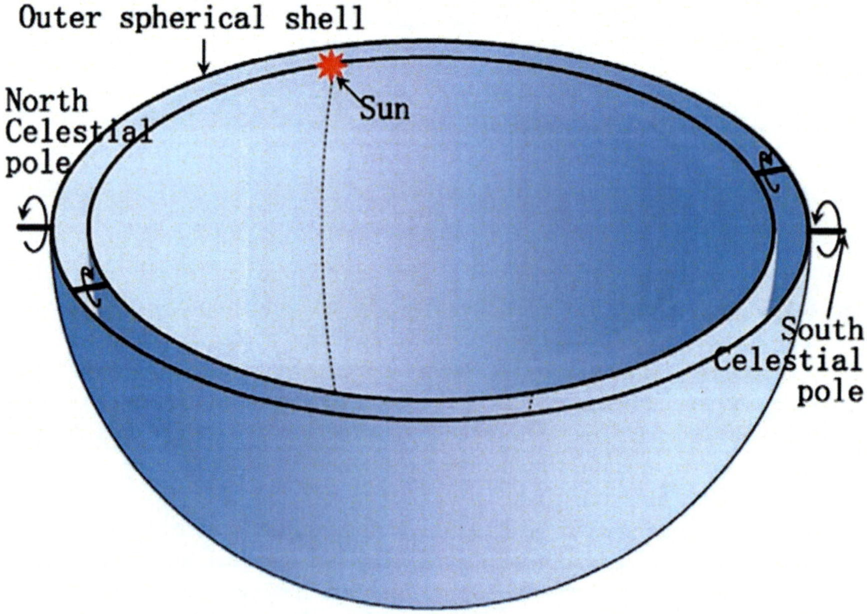

Fig. 4 Eudoxus's model of sun's motion

celestial sphere. The inner spherical shell is where the sun is located, and both ends of its axis are fixed on two opposite points of the outer sphere, forming an angle of 23.5° with the celestial axis, which is the obliquity between the ecliptic and celestial equator. The dotted line in Fig. 4 is the equator of the inner spherical shell (equidistant from both ends of the axis of the inner spherical shell), which becomes the ecliptic of the outer spherical shell, and the sun is on the equatorial line of the inner spherical shell to ensure that the sun moves on the ecliptic. The celestial sphere, together with the inner sphere, rotates around the spherical axis from east to west every day (a sidereal day). The inner sphere moves slowly from west to east and makes a circle a year while being forced to follow the celestial sphere to revolve. This is a reasonable explanation for the phenomenon that the sun rises in the east and sets in the west every day, and at the same time moves slowly along the ecliptic from west to east in the background of the celestial sphere.

It was not very hard for astronomers to think up this model because they had known for long that the ecliptic is intersected with the equator obliquely, and that the ecliptic pole is fixed at a certain point in the celestial sphere (actually it is not fixed, but its motion is too slow to be detected even for decades). This model can be said to be the direct result of observation, which is the combination of the celestial pole and the ecliptic pole. This model can be adjusted to also roughly describe the motion of the moon, but it did not solve the problem of the non-uniform motion of the sun and the moon.

The problem of the motion of the sun and the moon is barely solved, but it is hard for this model to demonstrate the motion of planets because of their retrograde motion. Therefore, Eudoxus added two more layers of spherical shells inside this model to simulate the retrograde motion of planets. The two new shells are combined as follows (Fig. 5), the planet is located at the equator of the innermost shell (the one with an inclined axis in Fig. 5), with its poles embedded in the second shell outside. The outer shell drives the inner shell to rotate. If the two shells rotate at the same speed but in opposite directions, and if the rotation axis of the two celestial spheres is not too different in direction, the trajectory of the planet will move up and down in a thin figure "8". The middle point of the figure "8" must be on the equator of the outer spherical shell, and its vertical centerline coincides with the right ascension of the outer spherical shell. The center of the figure "8" can be anywhere on the equator of the out shell, and the key is where the designer places the planet in the model in the beginning. The bigger the angle between the inner axis and the outer axis is, the bigger the figure "8" will be, otherwise, the smaller the angle is. The angle can be adjusted to adapt to the length of the retrograde distance of different planets.

Eudoxus needs to combine the two spherical shells in Figs. 4 and 5 together to demonstrate the motion of planets. Understanding this model requires excellent spatial imagination, and particularly interested readers can have a try. The general readers don't have to think it through, as long as they know that this combination can describe the retrograde phenomenon of the planet.

Figure 6 shows a planetary motion model composed of four layers of spherical shells. The planet is on the innermost layer, that is, the first layer of the shell. The two poles of the rotation axis of each spherical shell extend outwards and are implanted into the outer shell in turn, but the angle and position of implantation are critical.

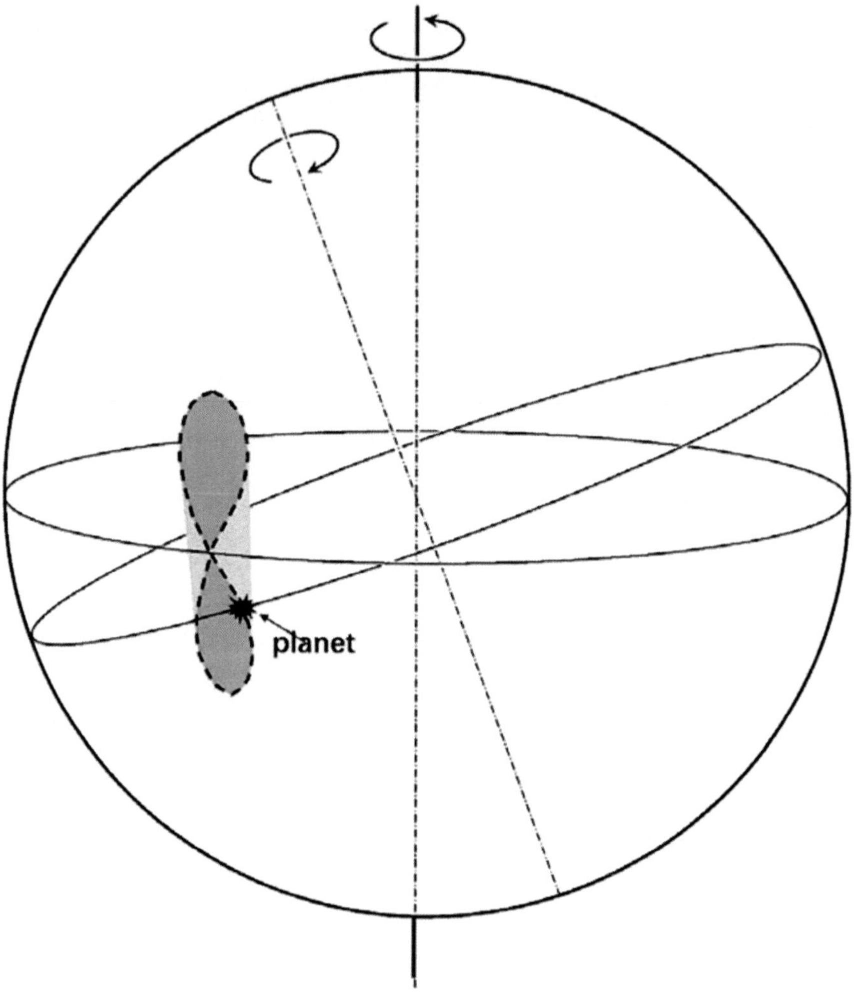

Fig. 5 Trajectory of the planet of the combination of two spherical shells

The axis of the outermost celestial sphere is the north–south celestial pole, which revolves around the earth uniformly from east to west, and it makes a circle around the celestial axis every day (a sidereal day). The revolution will be transmitted to the third, second and the first shell in turn, which can explain planets' rising in the east and falling in the west every day. The axis of the third shell inclines about 23.5° from the fourth to ensure that the planet moves near the ecliptic. The third shell revolves at a constant speed from west to east, and the revolving speed is the average speed of the planet moving in the ecliptic (such as Jupiter's 12-year cycle), so that the forward movement of the planet on the celestial sphere seen by people can be simulated. This revolution will naturally be transmitted to the second and first shells. The second

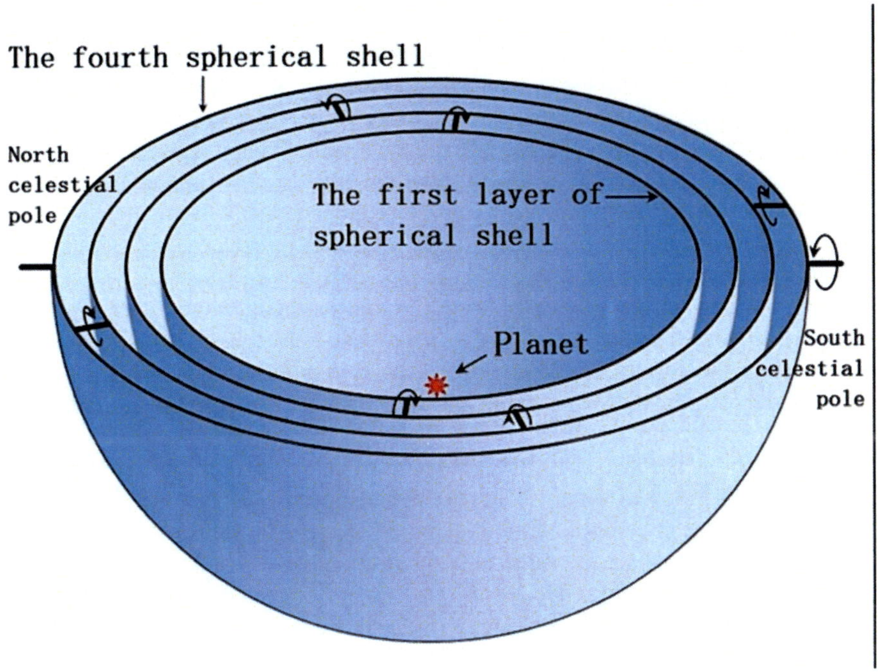

Fig. 6 Planetary motion model formed by four celestial spherical shells of Eudoxus

and the first shells are combined as shown in Fig. 5, which is then implanted into the third shell, but the rotation axis of the second shell must be perpendicular to that of the third shell, and the planets on the fourth shell must be adjusted to the position consistent with the ecliptic when they start moving.

Without the drive of the third spherical shell, the combination of the second and the first shell will control the planet to move in figure "8" (Fig. 5). However, driven by the motion of the third shell, the figure "8" is pulled apart, forming a trajectory similar to that is shown in Fig. 7, so the periodic retrograde motion of the planet is simulated. The axis angle and revolving speed of the spherical shell can be adjusted according to the different planetary movement cycles to roughly demonstrate the motion of each planet.

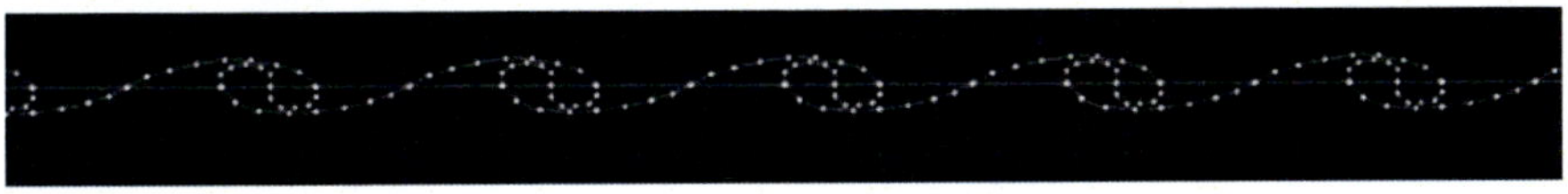

Fig. 7 Trajectory of planetary motion in concentric sphere model

This is really a complicated mechanical system, and it is not easy to come up with such a combination of circumference, which embodies Eudoxus's superb mathematical skills. He explained a variety of astronomical phenomena, especially the complex apparent motion of planets, with only uniform circular motion.

Eudoxus's concentric sphere is not a physical entity, but a theoretical aid. Moreover, the above model combined with spherical shells is only suitable for simulating one celestial body. There are seven groups of independent systems for the sun, the moon and the five planets, and they are not related to each other, so it cannot be called a system.

Aristotle, an ancient Greek philosopher and astronomer, accepted the idea of Eudoxus's concentric sphere, and he believed that these spheres existed. He wanted to integrate the seven planets (including the sun and the moon, which were also considered to be planets by ancient people), that is to say, the seven independent mechanical systems are integrated into a large and complete system where they contact with each other. Aristotle had to insert a total of 22 non-rotating spheres between each group of spheres carrying planets to prevent the unique motion of a celestial body from being directly transmitted to the celestial body below it. These non-rotating spheres are the same in number, rotation axis and speed as the sphere which planets move on above them, but they move in the opposite direction, to offset all the movement of the planet above and transmit only the diurnal motion to the planets below. This integration is rather far-fetched and it needs nearly 50 spheres in total, which shows that it is extremely complicated and incomprehensible without super spatial imagination.

Aristotle's system also adjusted the order of celestial bodies proposed by predecessors. The moon is the closest to the earth, followed by Mercury, Venus, Sun, Mars, Jupiter, Saturn and stars. Please note that the difference between this order and Plato's order lies in the positions of Mercury and Venus, which were ranked above the Sun by Plato while below the Sun by Aristotle. As for the reasons, they did not explain clearly.

Besides, there is a Primum Mobile outside stars in Aristotle's system. He believed that an object needs another object to push it to move, so he added a dynamic sphere, the Primum Mobile. The motion of the Primum Mobile is driven by the immobile gods, and once driven, it will transmit the motion to stars, the sun, the moon and planets one by one.

Aristotle said that the spheres in his system are as transparent as crystal balls, so they were often called crystal ball theory by later generations. These spheres, in fact, have to be transparent, or you can imagine how dark the universe will be if so many opaque spheres are combined. Therefore, many people really believed that there were several transparent crystal spherical shells in the sky for more than 1000 years.

Explaining the complex motion of planets with the combination of uniform circular motion is the biggest highlight of this model, which is undoubtedly a step towards the truth and an important enlightenment for later astronomers. However, this system was questioned in the beginning. The sun, the moon and planets should always keep a fixed distance from the earth according to this design, which is obviously inconsistent with the observation. Planets have distinct changes in brightness while moving, which means that planets are sometimes closer to the earth and sometimes farther away from the earth. When observing a solar eclipse, people also noticed that sometimes it was total and sometimes annular, which shows that the distance between the sun and the moon also changes, and it cannot be stimulated by the concentric sphere model. Similarly, this model cannot explain the non-uniform motion of the sun and the moon. In addition, this combination of multi-layer concentric spheres can only roughly demonstrate the retrograde curve of the planets, which is hard to match with the measured trajectory of the planets, and this can be seen from Fig. 7. Moreover, the model was not further developed because this kind of design is too complicated.

4 The Sun in the Center of the Universe

It will be more reliable to build a model of the universe if we can know the size and distance of celestial bodies before thinking about the model of the universe. How to measure their size and distance? Let's start with the sun, the moon and the earth under our feet, which we are most familiar with.

In fact, it is simple to measure the relative distance between the sun, the moon and the earth. The sun, the moon and the earth can be regarded as three points in the vast universe, and a triangle can be drawn by linking three non-collinear points. A triangle can be drawn almost any time because the sun, moon and earth are not in a straight line most of the time. Let's assume that the distance between the moon and the earth is 1, then as long as two angles of the triangle are measured, the triangle can be fixed. And it is simple to work out the relative lengths of the three sides. Even if we have not learned trigonometric functions, we can know the approximate number by drawing the triangle in proportion and measuring it with a ruler.

How to measure the three angles of this huge triangle? It should be easy to measure the included angle between the sun–earth line and the moon–earth line, but the other two angles must be measured on the moon or the sun. How is it possible? However, someone indeed "stood" on the distant moon by superb intelligence and measured the included angle. This man is Aristarchus who was born on Samos Island in ancient Greece.

When many astronomers in ancient Greece were pondering how the sun, the moon and stars revolve around the earth, he boldly put forward that it was the earth that moves around the sun, which was really a great shock. We know little about his life story, but he lived between 310 and 230 BC. He wrote a monograph *On the Size and Distance of the Sun and the Moon*, which showed us the power of deduction and reasoning of Greek geometry. It is necessary to introduce in detail the genius ideas of this outstanding astronomer in ancient Greece.

As shown in Fig. 8, illuminated by a parallel beam of light, half of a sphere is bright and the other half is dark. When an imaginary flat S is inserted into the boundary between light and dark of the sphere, it must be perpendicular to the parallel light.

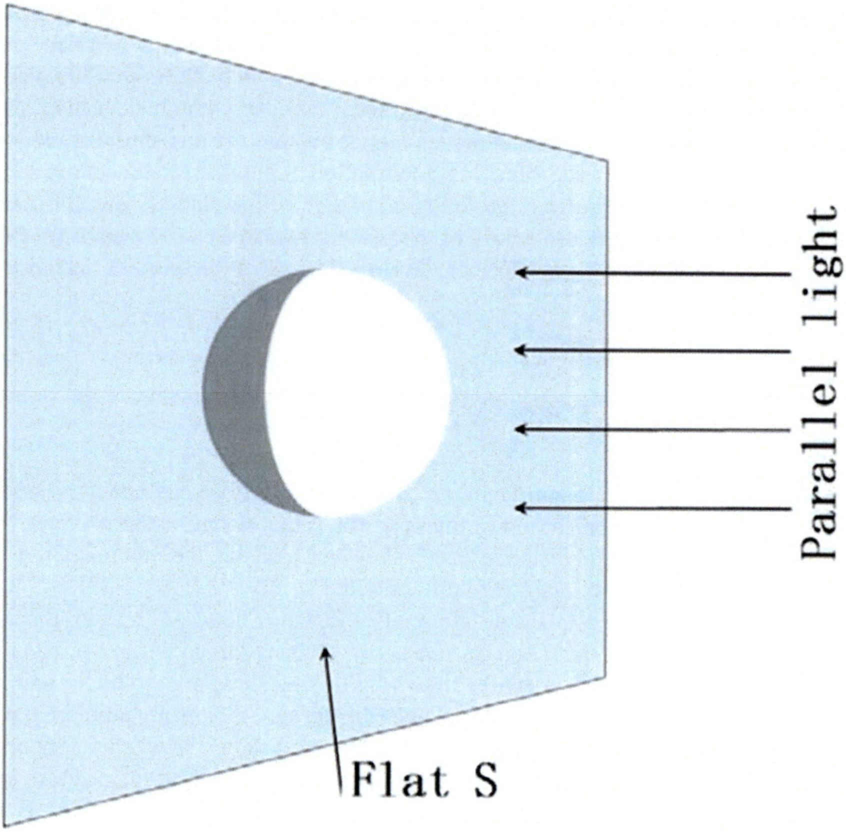

Fig. 8 The sphere is divided into light and dark halves under parallel light

Therefore, our line of sight should be perpendicular to the parallel light while looking at the center of the sphere at any point in flat S.

It has been testified by previous Greek astronomers that the moon's light comes from the sun. In addition, the distance between celestial bodies is very far, and the sunlight can be regarded as a parallel light approximately.

Aristarchus thus thought that observers should be on the imaginary flat S, with our line of sight and sunlight intersecting at the center of the moon at an angle of 90°, when the moon seen is half bright and half dark (this happens about the first quarter moon and last quarter moon). In this way, the line between the observer on the earth, the center of the moon and the center of the sun forms a right triangle (Fig. 9). This is the most critical step of this genius idea because we can measure the degree of the ∠moon in Fig. 9 at a specific moment by reasoning instead of standing on the moon. People who have learned geometry are familiar with the characteristics of right triangles, and the relative distance relationship of the sun, the moon and the

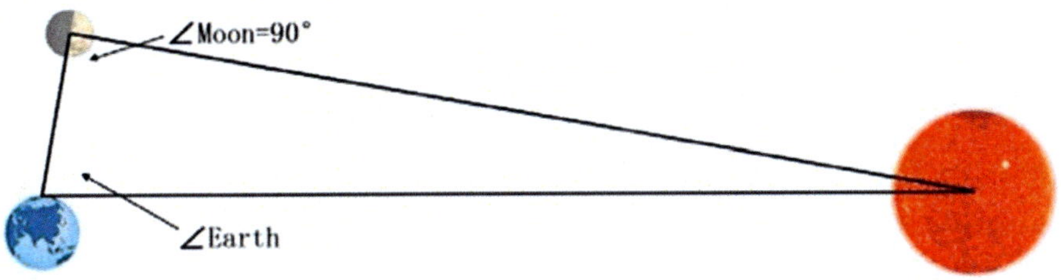

Fig. 9 A right triangle formed by the sun, the moon and the earth

earth can be calculated after measuring the included angle between the moon–earth line and sun–earth line (∠Earth in Fig. 9).

After measuring (it can be measured with instruments and the method of measurement should not be a problem, but accuracy is a big problem, which will be discussed later), Aristarchus got the result that ∠earth is larger than 87°. With this data, we know three angles and one side (the distance between the moon and the earth is set as 1) of the huge triangle, and it is not a difficulty to figure out the relative distance between the sun and the earth and the sun and the moon. Aristarchus calculated that the distance between the sun and the earth is more than 19 times that between the moon and the earth.

Seen from the earth, the sun and the moon are basically the same in size (Fig. 10). So, knowing the sun–earth distance is about 19 times longer than the moon–earth distance, the diameter of the sun is about 19 times longer than that of the moon, which can be calculated by using the simple proportional relationship of similar triangles.

Knowing the diameter of the sun is about 19 times longer than that of the moon, the following task is to work out the relative sizes of the sun, the moon and the earth. Aristarchus lived in an age when astronomers had already known that the lunar eclipse happened because the moon moves into the shadow of the earth. They also clearly knew that the moon moves about 13° from west to east in the sky every day, and it moves half a degree every hour, that is, it moves as far as one diameter of the moon per hour. By careful observation of the total lunar eclipse, especially the longest lunar eclipse (the longest means that the moon passes right through the center of the earth's shadow), the diameter ratio between the moon and the earth's shadow can be calculated. Aristarchus concluded that the diameter of the earth's shadow is slightly twice longer than that of the moon (Fig. 11).

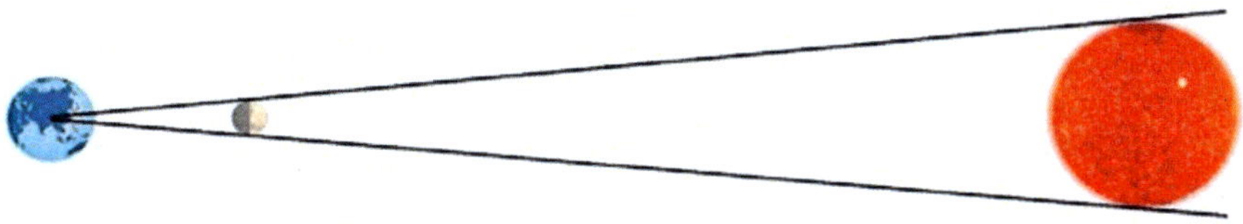

Fig. 10 Same angular diameters of the sun and the earth

Fig. 11 Diameter of the earth's shadow at the lunar eclipse

Please note that the earth's shadow here refers to the cross-section of the earth's shadow at the orbit that the moon is on during a total lunar eclipse, but what does the diameter of the shadow have to do with the diameter of the earth? Now we need to know the shape of the earth's shadow. The earth's umbra (the completely dark part that is not directly illuminated by the sun) will become bigger and bigger and in the shape of a trumpet if the sun is smaller than the earth. The earth's umbra will be cylindrical if the sun is as big as the earth. The earth's umbra will become smaller and smaller and in the shape of a cone if the sun is bigger than the earth. It is not hard to understand, is it?

According to the above triangle relationship, Aristarchus calculated that the diameter of the moon is 1/19 of that of the sun. The earth's diameter will be 19 times the moon's diameter if the sun is the same size as the earth, then it will take 18 or 19 h for the moon to pass through the earth's shadow during a total lunar eclipse, which is obviously inconsistent with the actual observation. There is no need for discussion if the sun is smaller than the earth, so the sun must be bigger than the earth and the earth's shadow must be conical.

So how much does the earth's cone-shaped shadow contract at the orbit of the moon? We know that the moon blocks the light of the sun from reaching the earth when it moves between the sun and the earth, which caused the total solar eclipse. The shadow of the moon must also be conical. The area on the earth where the total solar eclipse can be seen is very small whenever it happens, which indicated that the shadow of the moon is close to the tip of the conical shadow when it reaches the earth after such a long distance (right in Fig. 12), so it can be approximately regarded as the contracting diameter of a moon. Aristarchus believed that the moon moves in a circular orbit with the earth as the center. When the moon moves to the other side of the earth and enters into the conical shadow of the earth (left in Fig. 12), according to the principle of similarity, the diameter of the earth's shadow at this position can also be regarded as contract diameter of a moon, that is, the diameter of the shadow here = the diameter of the earth-the diameter of the moon. It has been demonstrated that "the diameter of the earth's shadow is twice that of the moon (Fig. 11)", so the earth's diameter is just over three times that of the moon.

Almost all the figures are available. The sun–earth distance is over 19 times greater than the moon–earth distance. The diameter of the sun is more than 19 times that of

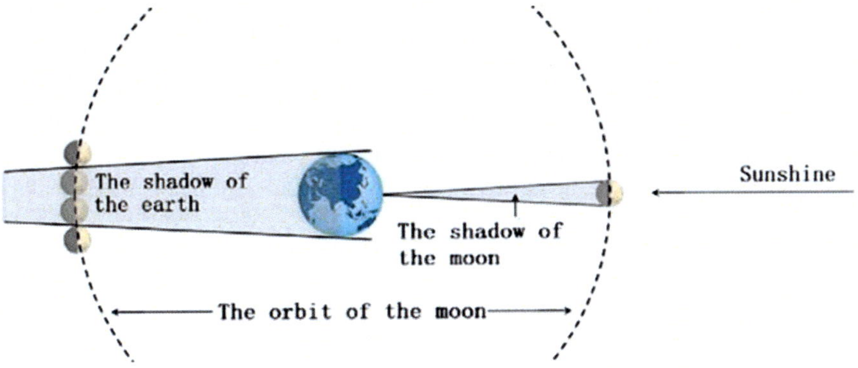

Fig. 12 Conical shadows of the moon and the earth

the moon. The diameter of the earth is about 3 times that of the moon. The diameter of the sun should be 6.75 times that of the earth, and the volume of the sun is more than 300 times that of the earth.

It is not complicated, is it? The knowledge of geometry is simple and clear, but the key is whether you can think of it. In people's eyes at that time, the earth was so huge, but the sun was more than 300 times bigger than the earth. How unimaginable!

We, living in the twenty-first century, have already known all kinds of real data about the sun, the moon and the earth. By comparison, Aristarchus's data is intolerably rough. For example, he calculated that the volume of the sun is more than 300 times that of the earth, but in fact, the volume of the sun is 1.3 million times that of the earth. What a great gap! But you know, that was more than 2000 years ago, in the third century BC, before the paper was invented.

People are used to the waxing and waning of the moon, and the total solar and lunar eclipses are not very rare, but who can imagine and reason like this? With just a pair of naked eyes, simple astronomical instruments and uncomplicated knowledge of geometry, Aristarchus was the first to measure the relative size and distance of the stars, which in itself, is a great achievement. We understand what a real genius is by knowing Aristarchus.

Using this method to measure the moon, the earth and the sun, we can only get very rough results, so the first difficulty, also the biggest problem, is how to accurately determine when the moon is semicircular. Even with a high-power telescope, we cannot do it, let alone with the naked eye, because the dark–light boundary of the moon is not a clear line in reality. People who have learned about triangle (secant function) know that, in the right triangle above, the closer the degree of $\angle$earth is to 90°, the bigger the ratio of the earth–sun distance to the moon–earth distance changes. For example, the sun–earth distance is 19 times of the moon–earth distance if the $\angle$earth is 87°, 29 times of the moon–earth distance if the $\angle$earth is 88°, and 57 times of the moon–earth distance if the $\angle$earth is 89°. The closer the $\angle$earth is to 90°, the greater the change is. It can be said that a minimal error or deviation may result in wide divergence. In fact, the real data of ∠earth is 89.6°, and the sun–earth distance is about 400 times of the moon–earth distance. What a great difference!

In spite of the rough result, Aristarchus's idea is clear, and the method is correct and excellent. The answer will be more accurate if the observation method can be improved. The practical value of geometric reasoning is that if a result is true in geometric reasoning, then it is still true (at least approximately true) in the real world. Genius Aristarchus proved that the sun is much bigger than the earth, which no one could refute after him. What the people later could do is to argue about how bigger the sun is than the earth, or to avoid talking about it.

The fact that a large celestial body is dominated by a small one is unreasonable and unbelievable, so Aristarchus had good reason to propose bolder ideas on the universe model. He pointed out that the stars and the sun are stationary, while the earth moves around the sun in a circular orbit, and the sun is in the center of orbit. He clearly knew that such a hypothesis would be challenged by others, just like the previous central fire model of Philolaus. If the earth moves around the sun that is stationary, it is conceivable that the earth's orbit is very large. Moving around the sun along such a huge circle, the earth will sometimes be closer to some stars, and sometimes farther away from them, and these stars should seem to expand or shrink when the earth gets close to or move away from them. But astronomers did not find this phenomenon, so Aristarchus thought that the earth must be constantly moving in the huge universe and the size of the earth's orbit is negligible compared with that of the celestial sphere. It implied that the vastness of the universe is unimaginable, and neither human beings nor the earth occupies a special position in nature.

Aristarchus's hypothesis is more than 1700 years earlier than Copernicus' heliocentric theory, and some people call him ancient Copernicus. It should be more correct to call Copernicus the Aristarchus of the sixteenth century.

Aristarchus's opinion is so advanced that other astronomers may not want to hear his "nonsense" at all except Archimedes, a great scientist of his time who paid close attention to and recorded his thoughts. People were more likely to accept Aristotle's view that heaven is superior to the earth in those days, because they could not believe that the earth under their feet was moving according to their intuition. If the earth is moving at a high speed, then won't everything on the ground, including people themselves, be thrown away? What's more, seeing is believing. People actually saw that it was the sun, the moon and stars that revolved around the earth!

Looking back at the above universe models, we can see how imaginative and intelligent some ancient astronomers were. Their hearts and minds had already flown away from the earth, even beyond the celestial sphere when they were guessing and imagining how the celestial bodies work.

Each of these universe models has its highlights and unique insights, containing reasonable elements. Whether they are affirmed or questioned, they are important inspirations and hints for later astronomers, and they are one solid footprint after another in mankind's exploration of the structure of the universe.

Chapter 6
Geocentric System Prevailing for a Thousand Years

The simple and harmonious central fire model cannot explain the speed and direction of planetary motion. The crystal ball model, which utilized the combination of circumference to simulate the retrograde motion of the planet, is complex and cannot explain the planets' changes in distance. The heliocentric model seems the closest to the truth nowadays, but few people believed it back then because it was ahead of time.

Just keep the earth at the center of the universe, as we cannot ask the astronomers to go beyond the times. As Plato said, the speed and direction of planetary motion is just a superficial phenomenon, and it can be explained by the combination of uniform circular motion. Astronomers keep on searching for a better combination to build a new model of the universe.

1 Magical Epicycle-Deferent System

Apollonius, an excellent mathematician and astronomer in ancient Greece at the end of the third century BC, put forward a new combination of circumference, that is, the epicycle-deferent System. According to his theory, the planets move uniformly in a small circle called an epicycle, whose center (point P in Fig. 1) moves along a larger circle called the deferent (Fig. 1).

This Epicycle-deferent System is so epochal that astronomers could not simulate the motion of the celestial body without it for over a thousand years after that, and the similar phenomenon of the system does exist in the actual universe, so we need to make a detailed explanation of this system to better understand the later models of the universe.

Many people have visited places like Happy Valley and amusement parks, where there is much large amusement equipment, such as roller coaster, rocker arm and so on, thrilling and unforgettable. Figure 2 is a schematic diagram of large amusement equipment, in which the main rotating arm revolves counterclockwise around the

Y. Ge, *From Geocentric to Heliocentric: How Discoveries Are Made*,
https://doi.org/10.1007/978-981-16-7779-3_6

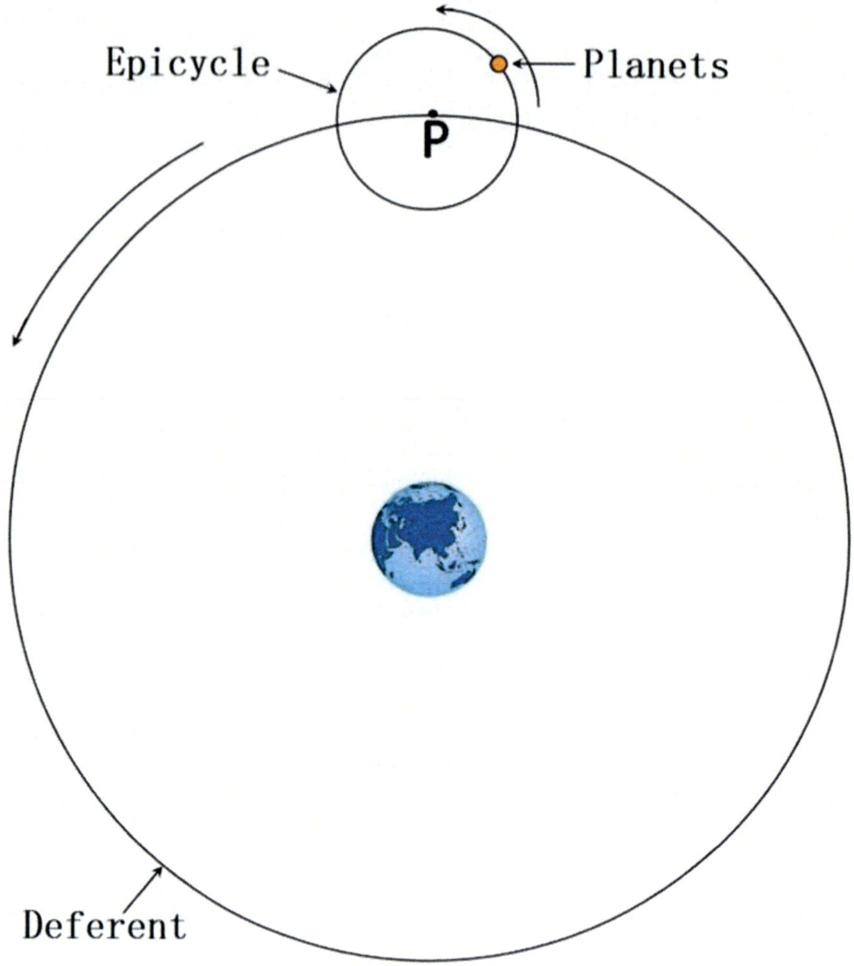

Fig. 1 Epicycle-deferent system

center O in the direction indicated by the arrow on the long-dashed line (later we call it forward rotation). Around the center S, the second rotating arm revolves counterclockwise in the direction indicated by the arrow on the dotted line (later we call it forward rotation), and it can also revolve clockwise in opposite direction (later we call it reverse rotation). To make it clear, three seats A, B, C are drawn in the picture, with only one tourist drawn, who is sitting at point B on the second rotating arm.

In this equipment, the tourist revolves around the center S, and the circle formed by revolution (dashed line in Fig. 2) is called an epicycle, and the center S is the center of the epicycle. What is the radius of the epicycle? It depends on where the tourist sits. If he sits in B, the radius of the epicycle is the distance from B to center S,

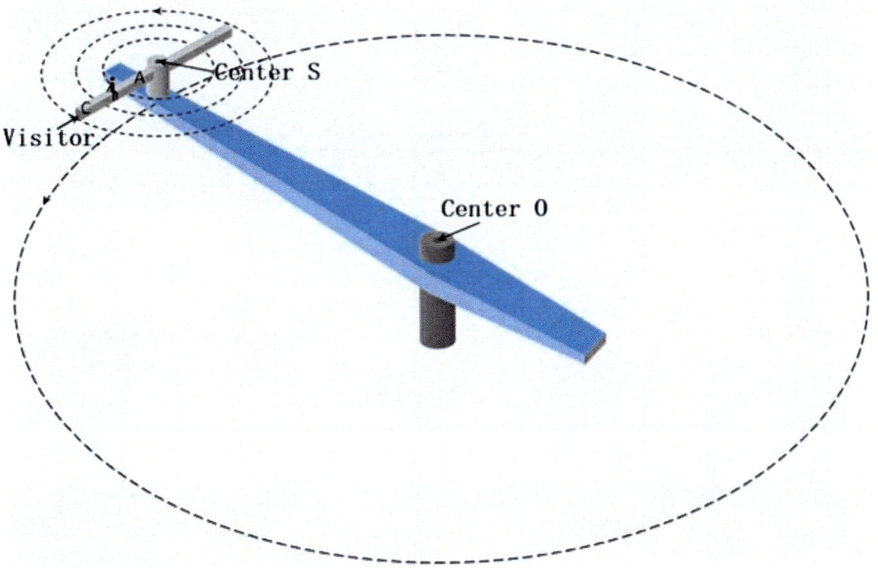

Fig. 2 An example of epicycle-deferent system

and the same applies to other positions. The circle (Long dashed line in Fig. 2) drawn by the center S in revolution around center O is called deferent, and the distance from the center S to center O is the radius of deferent. Let's assume that there is a high observation deck to the north of this large amusement equipment where many people stand, and how does the tourist move in their eyes? Several typical cases are selected for discussion, and the motions in the following figures are equivalent to those seen by people on the observation deck.

The first case is shown in Fig. 3. The secondary rotating arm will also make a circle in the same direction when the main rotating arm makes a circle (It can be seen from Fig. 3 that the tourist B is to the west of center S when the main rotating arm points to the north, to the south of center S when it points to the west, to the east of center S when it points to the south, and to the north of center S when it points to the east. When the main rotating arm points to the north again, tourist B returns to the west of center S, making a circle around center S). The combined movement is simple because the two movements are completely synchronized, and tourist B has made a circle around the center O at a uniform speed. If a person stands in center O, he will see tourist B revolve around him uniformly. The motion of the moon is similar to this, facing the earth with its same side. Seen from the earth, the moon only revolves around the earth, but does not rotate. But seen from the stars, it rotates once for every revolution around the earth.

The second case is shown in Fig. 4. The main rotating arm makes a circle and the secondary rotating arm keeps still (In Fig. 4, the tourist is always to the west of the center S). In eyes of people on the observation deck, the motion trajectory of the

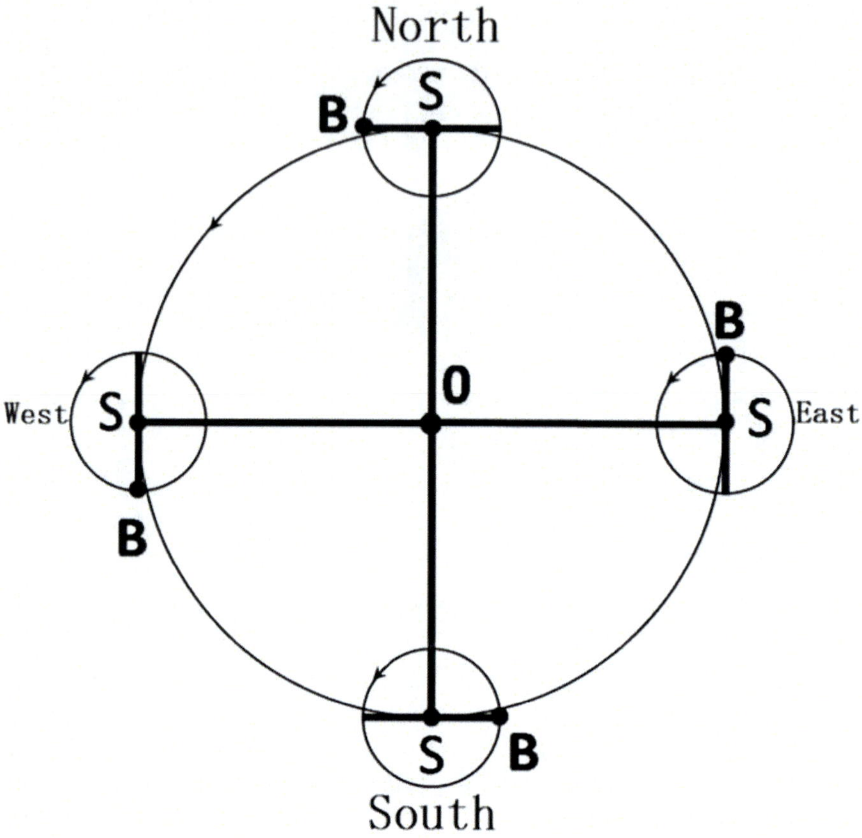

Fig. 3 The main rotating arm makes a circle and the secondary rotating arm makes a circle

tourist is still a complete circle (as shown by the dotted line in Fig. 4). The circle's radius equals the distance from the center S to center O, and its revolving speed is uniform, but its center is no longer on the point O. If a person stands at the center, he will see the distance between the tourist and him and motion speed of the tourist vary when the tourist circles around him, and the tourist moves faster when close to him and more slowly when far from him. The result of the combination of the two movements is equivalent to an eccentric circle, and the degree of eccentricity depends on the distance between the tourist and the center S.

The third case is shown in Fig. 5. The main rotating arm makes one circle and the secondary rotating arm makes two circles in the same direction. If the tourist sits at point A (a position closer to the center S) on the secondary rotating arm, and the distance between point A and center S is about 1/8 of the distance between center S and center O, then the people on the observation deck see the track diagram indicated by the dotted line in Fig. 6a, which is similar to a circle. If the tourist sits at point B and the distance between it and center S is about 1/3 of that between center S

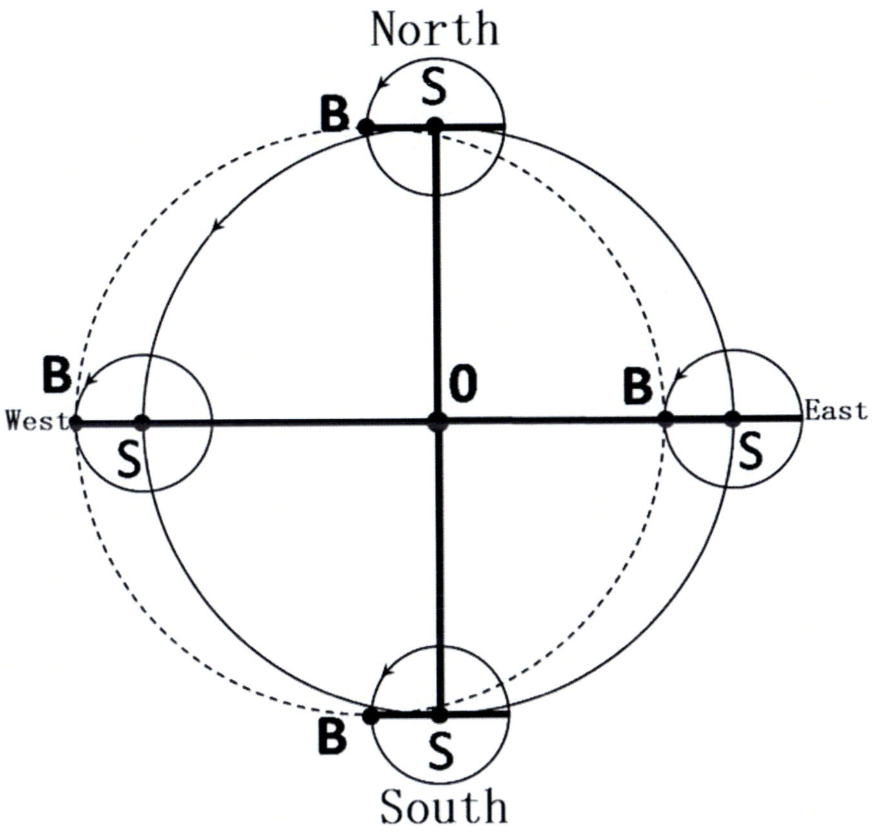

Fig. 4 The main rotating arm makes a circle and the secondary rotating arm keeps still

and center O, which means that the radius of the epicycle is bigger, then people see the track diagram indicated by the dotted line in Fig. 6b, which has been obviously deformed and cannot be called a circle. If the tourist sits at point C and the distance between it and center S is about 1/2 of that between center S and center O, which means that the radius of the epicycle becomes much bigger, then people see the track diagram indicated by the dotted line in Fig. 6c, which has a small gyration and the tourist moves backward in that small section. In addition, it can be seen from Fig. 6 that the speed of the tourist is not uniform. He moves faster when far from center O and more slowly when close to center O. In Fig. 6, the tourist's motion time between adjacent black spots is equal, so he moves faster when the adjacent distance is long and more slowly when the adjacent distance is short.

The fourth case is shown in Fig. 7. The main rotating arm makes one circle and the secondary rotating arm makes six circles, that is to say, the secondary rotating arm revolves faster and faster, then what is the trajectory of the tourist like? Figure 7 shows trajectories of the tourist at points A, B and C, and we can see five beautiful

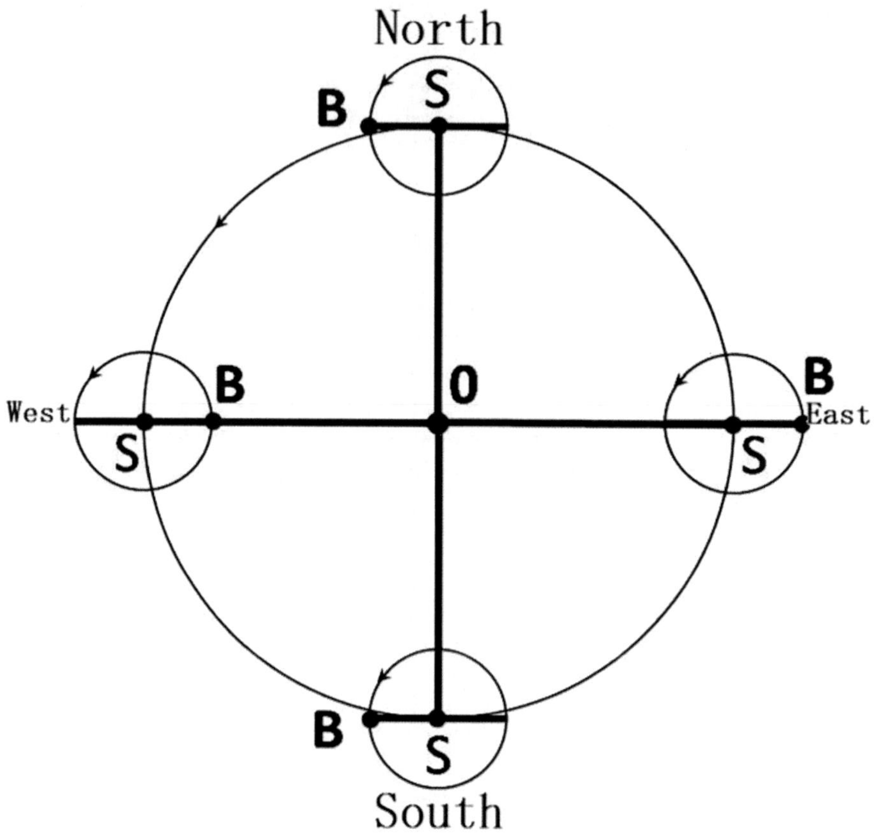

Fig. 5 The main rotating arm makes one circle and the secondary rotating arm makes two circles

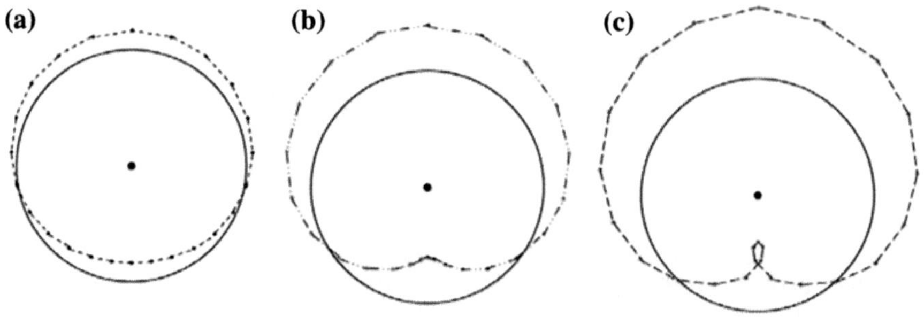

Fig. 6 The tourist's motion trajectory in different positions when the main rotating arm makes one circle and the secondary rotating arm makes two circles

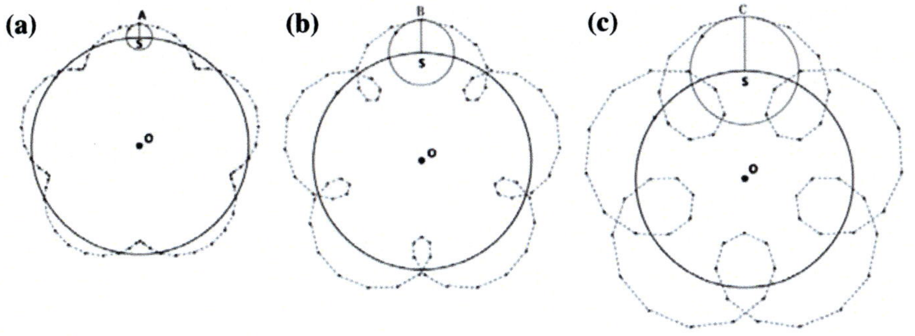

Fig. 7 The tourist's motion trajectory in different positions when the main rotating arm makes one circle and the secondary rotating arm makes six circles

petals in each picture. The petal is small when the epicycle is relatively small and the petal is big when the epicycle is relatively big. If someone stands at center O, he will see that when the tourist is at point A, there are only changes in speed and direction, no retrograde motion; when the tourist is at point B, he moves backward five times, but the retrograde range is small; and when the tourist is at point C, he also moves backward five times, and the retrograde range is big.

We have described four situations, and some regularities can be derived. First of all, the tourist moves uniformly on the circle all along, and the tourist's trajectory seen by people on the observation deck is a real trajectory composed of two circles. Secondly, the trajectory pattern of a tourist's motion depends on the radius ratio and speed ratio between epicycle and deferent. If we change the radius and speed, but do not change the proportional relationship between them, then the trajectory pattern of tourist's movement is the same as the original, and the law of speed change remains the same; if we change their proportional relationship, making the epicycle larger and revolve faster, then the tourist may move backward. The larger the epicycle is, the greater the range of retrograde motion of tourist will be; and the faster the epicycle rotates, the more retrograde motions of the tourist will appear in the process of making a circle with the main rotating arm. Combining Figs. 6 with 7, we know that if the main rotating arm makes one circle and the secondary rotating arm makes ten circles in the same direction, then the tourist's trajectory will draw nine beautiful petals. The rest can be done in the same manner.

For people standing at center O to watch the tourist's motion, there are changes in distance, speed and even the alternation of moving forward and moving backward according to different radius ratios and speed ratios. In other words, by properly changing the relative sizes and speed of the epicycle and deferent, this synthetic circular motion system can describe all kinds of motion, far or near, fast or slow, as well as forward or backward. Thus, we can see that there are countless ways to combine epicycle and deferent, which are versatile, and it works well to simulate motions of planets with this combination.

We have no idea how Apollonius came up with such a plan, but we can have a guess. Maybe he was inspired by studying the laws of Mercury and Venus. These two stars always move back and forth around the sun, and obviously, they do not revolve directly around the earth. Perhaps as Heraclitus of Plato School said, they revolve around the sun and then revolve around the earth with the sun; or perhaps they revolve around some point on the sun-earth line and then revolve around the earth with this point. Only in this way will they always be as close to the sun as we have observed, and it is a living epicycle-deferent model, isn't it? Such a model is shown in Fig. 3 in Chap. 5.

Mercury and Venus move in their own epicycle, while the center of their epicycle revolves around the earth. So, will Mars, Jupiter and Saturn also have their own epicycle and deferent? According to observation, these planets are sometimes very close to us, sometimes far away from us, sometimes fast, sometimes slow. They move forward from west to east most of the time, but they also move backward regularly. It can thus be seen that they do not revolve directly around the earth, and maybe they also have their own epicycle and deferent like Venus and Mercury. By describing the motion of Mars, Jupiter and Saturn with the epicycle and deferent, their changes in motion direction, speed and distance from us are consistent with the actual observation. The motion trajectory of Jupiter simulated by the combination of epicycle and deferent is shown in Fig. 8.

Due to its ingenious design, strong adaptability and powerful function, the combination of epicycle and deferent is second to none in the history of science until modern times. In hindsight, the introduction of epicycle and deferent has put Greek astronomy on the right track, and it is very encouraging that the complex motions of the five planets can be simulated in a highly approximate way with the combination of epicycle and deferent.

At the same time, it should be noted that there is another important method about the combination of epicycle and deferent which has not been described, because in more than one thousand years after Apollonius found the combination of epicycle and deferent, no astronomer or mathematician, including Apollonius himself, noticed that there was another combination. This combination method is also significant, and we will discuss it later.

2 Let the Earth Deviate Slightly from the Center

Some astronomers say that Apollonius also put forward another plan, namely, the eccentric circle system, but some say that the eccentric circle system was proposed by the later astronomer Hipparchus. Actually, the epicycle-deferent system includes the eccentric circle system. Figure 4 shows the eccentric composed by epicycle and deferent, and by adjusting the size of the epicycle, the degree of deviation of the eccentric circle can be adjusted. Apollonius, however, did not specify what planet corresponds to what eccentric circle, and part of this task was done by Hipparchus.

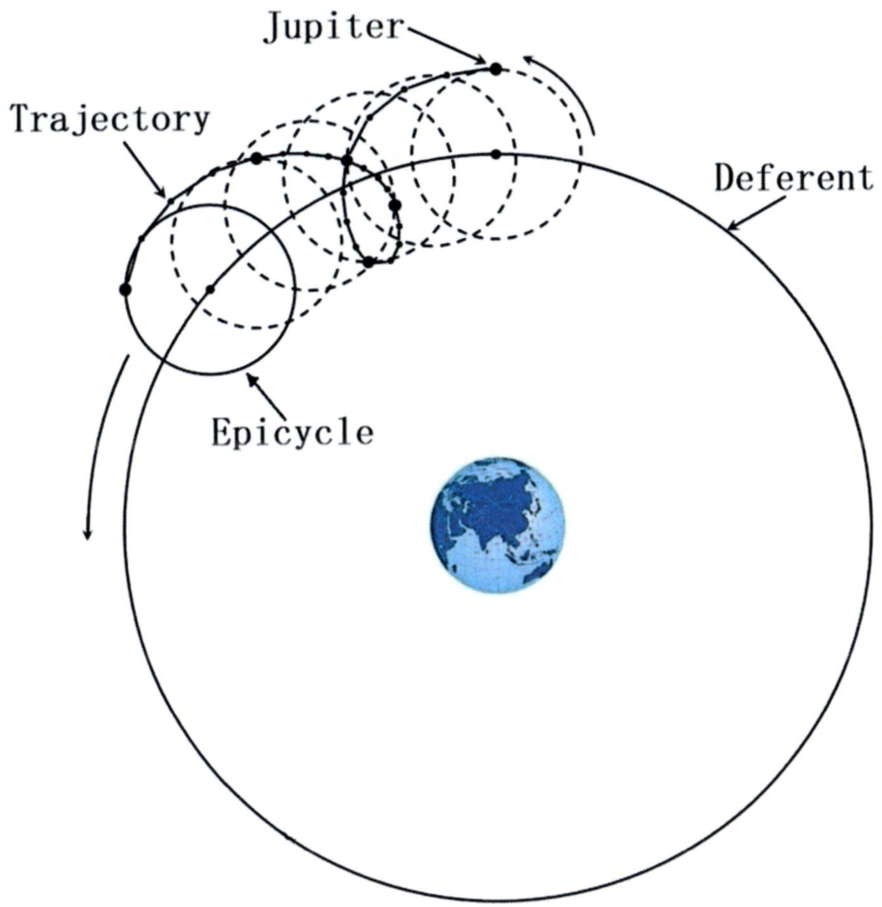

Fig. 8 The motion trajectory of Jupiter simulated by combination of epicycle and deferent

Hipparchus is a great Greek astronomer whose works have not survived due to age, and what we know about him today is from Ptolemy's works. We do not know about his life, and the only information is that he lived in the second century BC. He has made many significant contributions to astronomy, which we will not list one by one, but will give just two examples to appreciate the vision and wisdom of this genius.

As has been mentioned before, we have to count if we want to know how many days there are in a month, such as counting from one full moon to the next full moon. When is the full moon? Some people may say that it is the time when the moon is full, but when is it? With the naked eye alone, it is not easy to determine the day of the full moon, let alone the exact moment of the full moon, because the moon looks similar on the 15th day and 16th day in a month. Astronomers certainly know that as long as the right ascension of the sun and the moon are carefully measured, it will be the exact moment of the full moon when they are 180° apart. Can we start counting when we have the exact moment? Yes, but one can count for 30 or 40 years at most in his life, and it is hard to get more accurate data. We know that the longer the counting time is, the more accurate the average value is, which applies to all periodic motions. Astronomers can count from generation to generation to get more accurate values, and how about counting for three or four hundred years? The data should be accurate enough then. But this method is too troublesome and too unwise.

By the time Hipparchus lived, astronomers had already understood the reasons for the lunar eclipse. The earth blocks the light from the sun to the moon when they three are completely in a straight line, so the lunar eclipse occurs. The moment of lunar eclipse is the exact moment of the full moon, which is not difficult to understand, and the number of months between any two lunar eclipses must be an integer, so isn't it good to count from one eclipse to the next one? How come no one thought of this before?

Hipparchus compared his accurate eclipse observation records with those handed down from Babylon, and chose two representative eclipses. The two eclipses are separated by 4267 months, which is a total of 345 Egyptian years (365 days per year in the ancient Egyptian calendar) plus 82 days and one hour, that is, 126,007 days and one hour. Dividing 126007.04 days by 4267 months, we will get an average of 29.53059 days per month. Compared with accurate value today, it only has a slight error in the fifth decimal place. How clever this method is!

A total solar eclipse occurred in central China on July, 22, 2009, which can be seen in Anqing City and Tongling City in Anhui Province that was in the center of the eclipse zone. At about the same time, a partial solar eclipse was seen in Beijing City, the due north of Anqing and Tongling, and approximately 4/5 of the diameter of the sun was blocked when the moon blocked the sun the most.

Do these data mean anything to you? Do you think of anything from this? In the second century BC, Hipparchus also met a total solar eclipse. One day, a total solar eclipse occurred near Turkey, and at the same time in Alexandria, Egypt, a partial solar eclipse was seen, and the diameter of the sun was covered by 4/5 of the maximum. As an astronomer, Hipparchus keenly captured this important information, because it was equivalent to measuring the parallax of the moon, although the measurement was rough.

What is parallax? Here we need to explain the concept and application of parallax, which is often mentioned in astronomy.

A man standing on the south bank of a river wants to measure how far the pole on the other side of the river is from him. What should he do? As shown in Fig. 9, one is standing at point A to observe a pole D on the other side of the river, and he feels that the tree behind the pole is on the east side of the pole. When he observes at point B, he sees that the tree is on the west side of the pole. The change in apparent positions of the tree and pole is called parallax. We just need to measure the distance between point A and point B and the degrees of angles A and B in Fig. 9 to measure the distance from point A to point D, instead of going to the other side of the river. Draw a triangle completely similar to the actual situation based on these three data, the distance between point A (the same with point B) and point D can be calculated

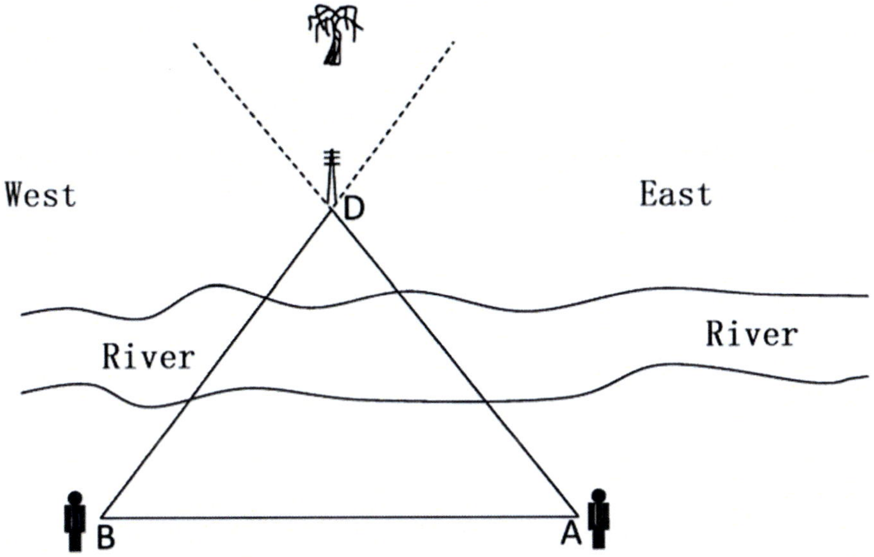

Fig. 9 Parallax ranging method

according to the characteristics of the triangle. The celestial bodies are so far from us that we cannot measure them directly. It is the geometry that provides us with a solution, enabling us to fly thousands of miles to measure the distance between the celestial bodies and us. This is the parallax-ranging method often used in astronomy, and this method will often be mentioned later.

Knowing the parallax of the moon, Hipparchus could calculate the relative distance between the moon and the earth. It is said that the process of proof and calculation is as shown below, and interested readers may have a look and try.

As shown in Fig. 10, point B is the Hellespont near Turkey, where a total solar eclipse was seen, and point A is Alexandria in Egypt, where a partial solar eclipse was seen with 4/5 of the diameter of the sun blocked in maximum. The angular diameter of the sun is 0.5°, and 1/5 of the sun is seen at point A (CD in Fig. 10), so the $\angle CAD$ is 0.1°. It can be seen from Fig. 10 that $\angle BEA = \angle EDA + \angle EAD$ (exterior angle of the triangle is equal to the sum of two nonadjacent interior angles), and because the sun is much farther from us than the moon (the ancient Greeks thought it was about 20 times, which is actually nearly 400 times, and the actual scale cannot be drawn in Fig. 10), $\angle EDA$ is much smaller than $\angle EAD$. Therefore, $\angle BEA$ can be regarded as approximately equal to $\angle EAD$, which is equal to 0.1°. EA and EB can also be approximately equal because the moon-earth distance is extremely far.

In this way, the triangle EBA is an isosceles triangle, and we know that its vertex angle is equal to 0.1°. So we can calculate the length of EA (or EB) as long as we know the distance between A and B. Hipparchus calculated that the distance between Hellespont and Alexandria is about 1/40 of the circumference of the earth according

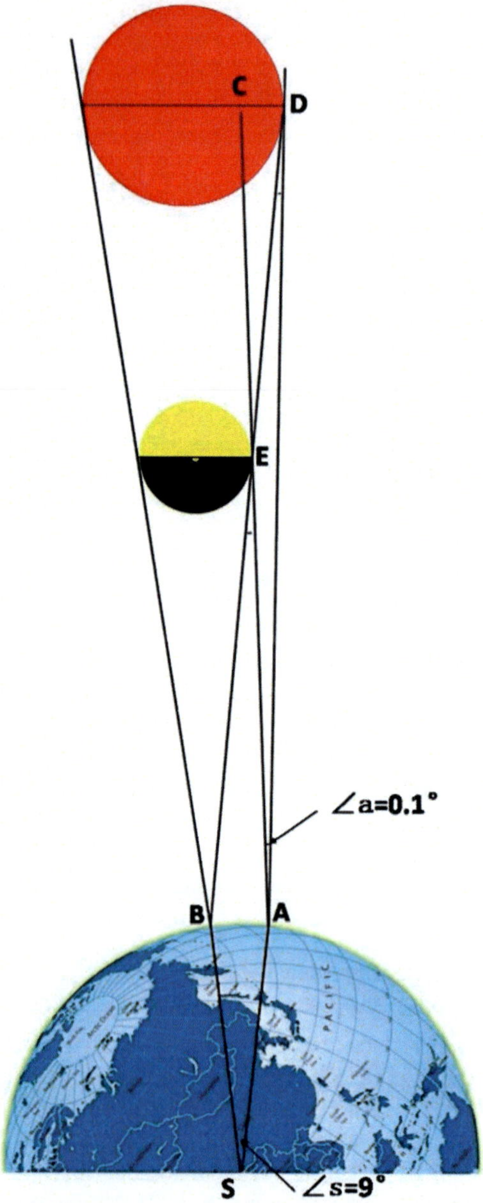

Fig. 10 Measurement of the moon-earth distance according to a lunar eclipse

to the data of longitude and latitude of the two places, so the distance between the moon and the earth is about 90 times the radius of the earth.

This kind of measurement and calculation is rough, but the vision is genius. It is said that Hipparchus later perfected the data and recalculated that the moon-earth distance is about 59–67.3 times the radius of the earth.

After that, by observing the altitude of the moon at two different latitudes, Hipparchus calculated that the distance between the moon and the earth is about 30 and 1/6 times the diameter of the earth, which is slightly smaller than the actual value.

Through these two examples, we can see the unique vision and extraordinary wisdom of an outstanding astronomer.

Hipparchus did not give a systematic model to explain the structure of the universe, but he explained the motion of the sun with an eccentric circle carefully, and gave the concrete value of this model.

We have known that it takes over 365 days for the sun to travel along the ecliptic from west to east on the celestial sphere, and that the great circle of the ecliptic is divided into four sections by the vernal equinox, the summer solstice, the autumn equinox and the winter solstice. The angular distance between two adjacent points is all 90°, so it seems that the ecliptic circle is divided into four equal parts. However, it is found that, after careful observation, it takes 94.5 days for the sun to travel from vernal equinox to summer solstice, 92.5 days from summer solstice to autumn equinox, 88.125 days from autumn equinox to winter solstice, and 90.125 days from winter solstice to vernal equinox. Doesn't this show that the sun does not move at a uniform speed in the sky?

By observation and measuring, we can find that the apparent diameter of the sun changes in a year, so the distance between the sun and the earth also changes. How to explain these phenomena?

According to Hipparchus, the earth is not in the center of the sun's circular orbit, but eccentric, which is the famous eccentric circle plan. Of course, it is not enough that we simply know it is eccentric, we also need to know which direction it deviates towards and how much it deviates. Hipparchus, after calculation, pointed out that the earth should deviate from the radius of the sun's circular orbit by 1/24, and the angle between the line connecting the earth and the center of the sun's orbit and the line connecting the earth and the vernal equinox should be 65.5°. Figure 11 is the eccentric circle model of the sun's motion of Hipparchus. We can see that the sun, seen from the earth, moves a little longer in the upper half than in the lower half because the earth is not in the center of the circle. Therefore, although the sun moves uniformly on the circle, it is not uniform seen from the earth.

Just move the earth a little bit away from the center of the circle, and the problem can be solved nicely. Does it seem very simple to you? The eccentric circle plan of Hipparchus does not contradict the old rule that the celestial body moves uniformly in the circle, but it also well explains the non-uniformity of the sun's apparent motion. After Hipparchus, the study of the universe model is no longer confined to the construction of a general and rough outline, but enters the stage of quantitative, data-supported prediction of the position of celestial bodies. Therefore, the

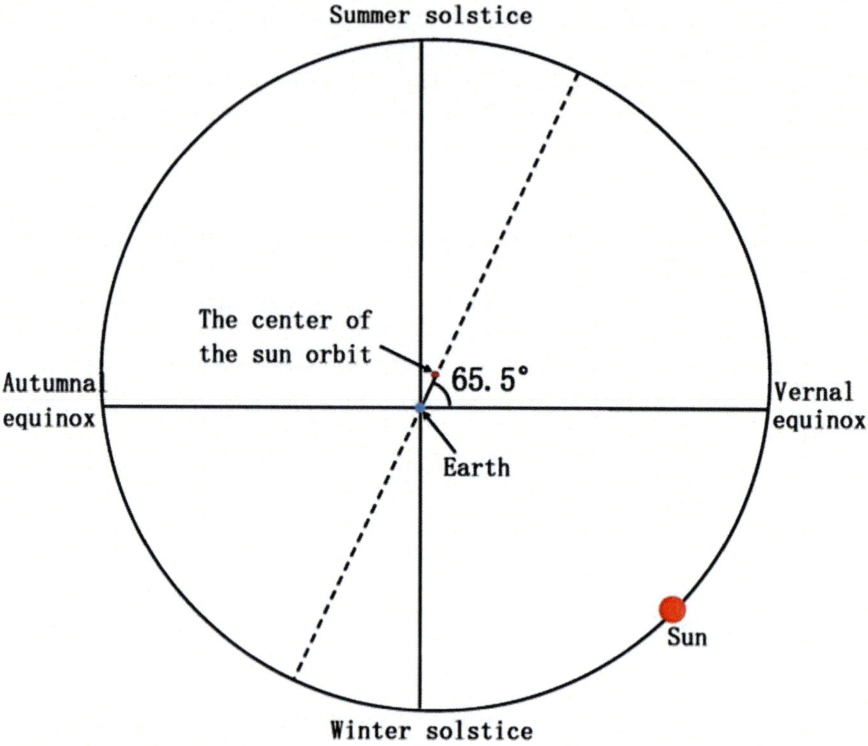

Fig. 11 The eccentric circle model of sun's motion by Hipparchus

illustration later will involve some specific data, which is a little "profound", but it is not hard to understand with a little patience.

3 A Complete Geocentric Model

The splendid work of Apollonius and Hipparchus laid a solid foundation for Ptolemy's research.

There is little historical information about Ptolemy's life. His ancestors were probably Greeks or Hellenized people of a certain race. He was active during the reign of Roman emperors Hadrian (reigned 117–138 AD) and Antoninus (reigned 138–161 AD) and lived until the reign of emperor Marcus Aurelius (reigned 161–180 AD). He made all observations in Alexandria, Egypt (then under the rule of the Roman Empire). It is said that he lived near a very large library with many neighbors who were scholars, enjoyed a good academic atmosphere, and was deeply influenced by Plato and Hipparchus. He is remembered because of his academic achievements. He has more than a dozen works handed down from ancient times, among which the most famous is *Almagest*. It is an encyclopedia of astronomy, and it has been the highest authority in astronomy for over one thousand years since its publication.

In his book, Ptolemy constructed a complete geometric model to describe various movements of celestial bodies such as the sun, the moon, the five planets and all stars. He imagined that there are nine heavens in the universe, that is, nine revolving concentric crystal spherical shells. The moon is the lowest, followed by Mercury and Venus. The sun is on the fourth celestial sphere, which is the king of the universe and soul of the world, illuminating the universe with its great brilliance. The fifth to seventh heaven are Mars, Jupiter and Saturn in turn. The eighth heaven is the sphere of fixed stars, and all the stars are embedded in the layer of these heaven-like gems. Above the sphere of fixed stars is the highest celestial sphere, that is, the Primum Mobile, where the gods live. Each layer of the spherical shell is thick enough to hold all the epicycle, and one connects the other. The farthest distance between a celestial body and the earth is equal to the shortest distance between the celestial body above and the earth. All spheres are driven by the Primum Mobile and make diurnal motion around the earth from east to west. Except for the sphere of fixed stars, all the other seven heavens have their own motions that are opposite to the diurnal motion (Fig. 12 is a plan sketch, and Fig. 13 is a three-dimensional diagram drawn by Europeans).

There seems no novelty about this system, and the arrangement of the sun, the moon and planets has not changed for hundreds of years except Mercury and Venus, which are always controversial. The difference is that Ptolemy used the epicycle-deferent system of Apollonius and the eccentric circle system of Hipparchus to explain the motion of the sun, the moon and the planets.

Most of the universe models before Ptolemy's emphasized the qualitative description of the universe structure. What they emphasized was creativity, conception, and the general motion of planets, just like a draft or outline. They did not demonstrate the details of motions of celestial bodies, and they paid little attention to the calculation accuracy of the specific celestial body's movement in their universe model. In addition, they did not provide the position of each celestial body at any time. However, without the support of details and detailed drawings, it is hard for other astronomers and the public to accept it. If the details derived from a certain kind of creativity and conception are inconsistent with the real celestial phenomena, then the creativity and conception are doubtful. On the contrary, it is convincing if the details derived from a certain kind of creativity and conception are highly consistent with the real celestial phenomena. A universe model only used to explain phenomena was not enough for Ptolemy, he hoped that his model could not only explain the phenomena well, but also be applied to make accurate calculations.

Ptolemy thus carried out complicated and arduous work. What is the ratio of the epicycle to the deferent of each celestial body? What is the moving direction and speed of the celestial body on the epicycle? What is the moving direction and speed of the center of the epicycle on the deferent? For every celestial body, how far does the earth deviate from its orbit center, and which direction is the earth in? Ptolemy synthesized the observation data of predecessors and his own, gave the specific parameters one by one through complex calculation, and finally compiled various astronomical tables, so that he could calculate the positions of various celestial bodies

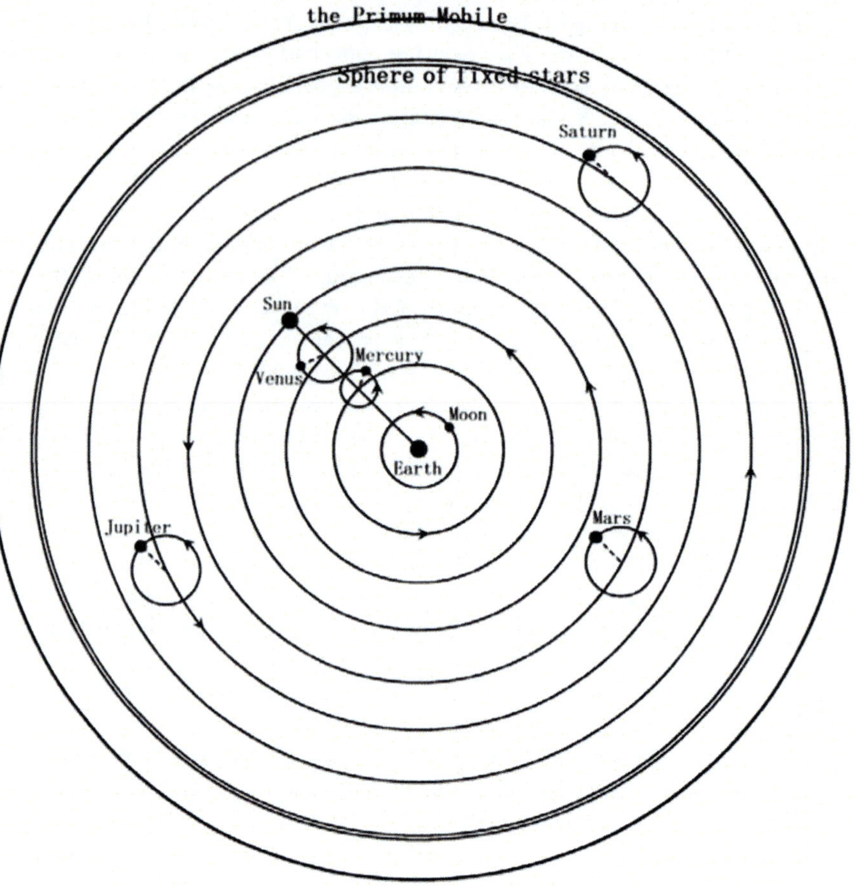

Fig. 12 Ptolemy's geocentric model

in advance at any given moment. It is significant that the positions of celestial bodies calculated according to Ptolemy's model are very close to the actual observation, which is the fundamental reason why Ptolemy's geocentric system can be called a system, and the greatness of this system lies in this.

How did Ptolemy determine the speed of the epicycle and deferent of the planets, the sun and the moon?

Astronomers knew long before that it takes 29.5 years for Saturn to orbit from west to east in the background of the celestial sphere, 11.86 years for Jupiter, 1.88 years for Mars, and 1 year for Venus and Mercury. Therefore, the respective orbital period (also the orbital period of the rotating arm in the analogy above) of them must be 29.5 years, 11.86 years, 1.88 years, 1 year and 1 year.

The difficulty lies in measuring the revolving speed of each planet on its own epicycle, that is, the speed of the secondary rotating arm in the analogy above. Figure 14 is a schematic diagram of the relationship between Jupiter's epicycle and

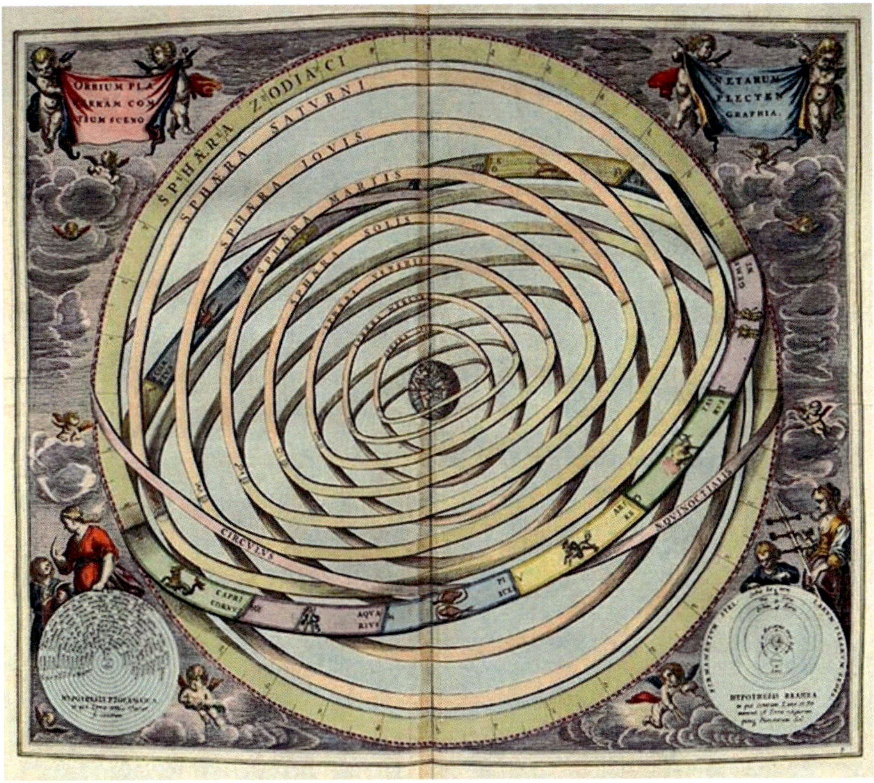

Fig. 13 Stereoscopic diagram of Ptolemaic system drawn by European (The picture is from the Internet)

the sun's orbit, and after observation, it takes 11.86 years for Jupiter's epicycle to make a circle on the deferent. Let's choose a time when Jupiter and the sun are on opposite sides of the earth. Jupiter is at point A1 of the epicycle and the sun is at point S1 when the center of Jupiter's epicycle reaches point P1, and at the same time, Jupiter and the sun are in opposition when Jupiter is the closest to the earth and the brightest. After 399 days, the sun has made a little more than a circle in its own orbit, reaching point S2, and the epicycle of Jupiter reaches point P2, so Jupiter and the sun are on opposite sides of the earth again. How fast should Jupiter move to make itself reach point A2 and be opposite to the sun? It can be seen from Fig. 14 that Jupiter's angular velocity on its epicycle must be exactly the same as that of the sun in orbit, that is, one revolution a year, so that this requirement can be met. In this way, Jupiter is certain to move backward when it is closest to the earth, which is consistent with the observation. That is exactly how Ptolemy determined the speed of Jupiter on its epicycle.

It takes 11.86 years for Jupiter's deferent to make a circle and 1 year for its epicycle, and for us, who are standing in the center of the deferent, it appears that

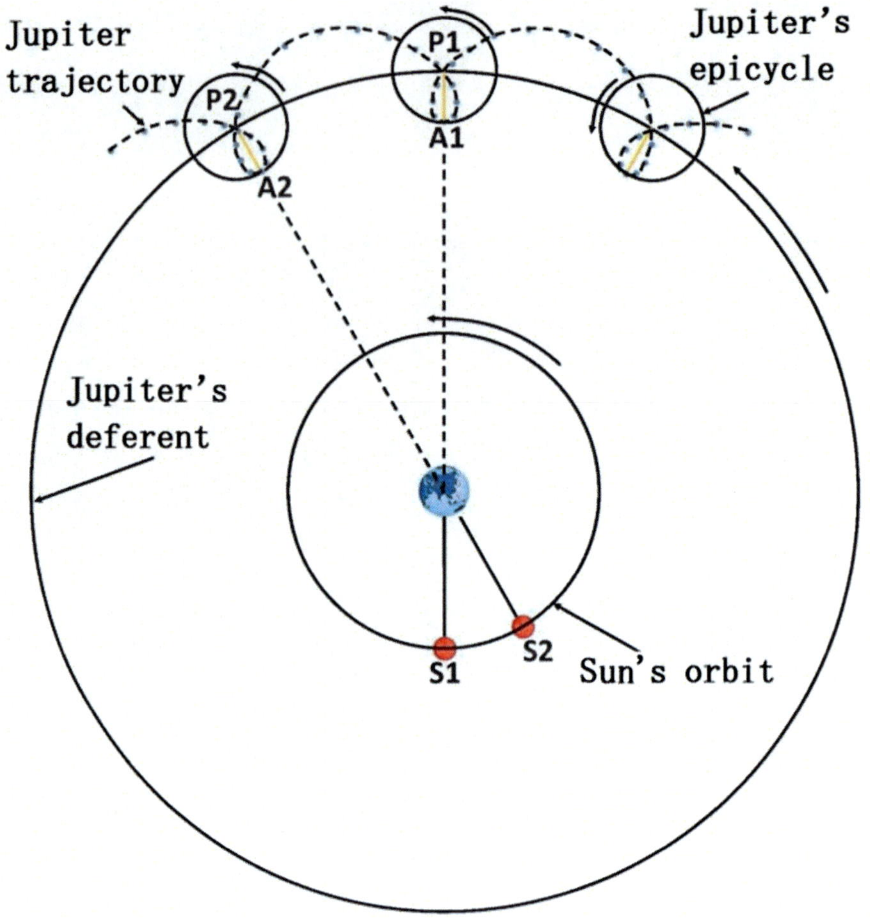

Fig. 14 The diagram of the relationship between the Jupiter's epicycle and the sun's orbit

Jupiter has an average of nearly 11 retrograde motions in 11.86 years. Figure 15 is a schematic diagram of Jupiter's trajectory. We see that Jupiter has drawn 11 beautiful petals in 11.86 years.

In the same way, the time for Saturn to make a circle on its epicycle must be one year, and the time for Mars to make a circle on its epicycle must be one year, too. Therefore, in Ptolemy's universe model (Fig. 12), we see that the lines between Mars, Jupiter and Saturn and centers of their respective epicycle are always parallel to that between the sun and the earth.

So what about Venus and Mercury? Since they are always around the sun, which means that the center of their epicycle must be on the sun-earth line, it takes a whole year for their epicycle to make a circle on their respective deferent. Astronomers have long known that Venus is in superior conjunction (or inferior conjunction) with the sun for five times in 8 years, and seen from the earth, Venus has turned around its

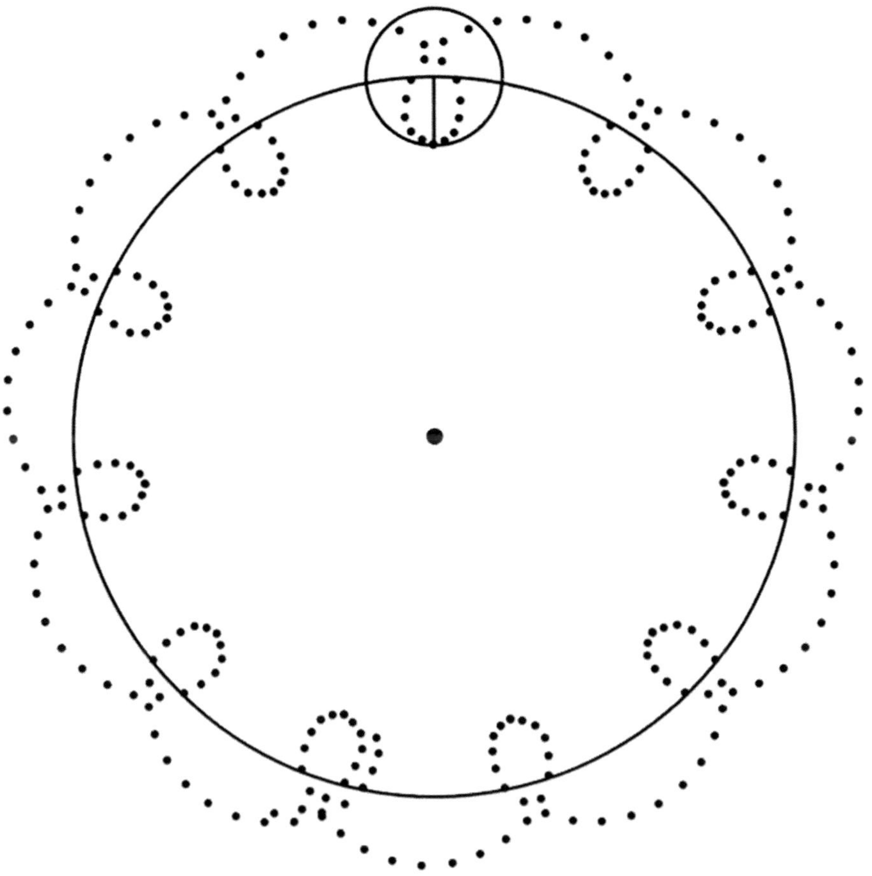

Fig. 15 Jupiter's trajectory

epicycle five times. The center of Venus's epicycle, like the sun, has made 8 circles around the earth in 8 years. As a result, Venus has made 13 (= 5 + 8) circles around the center of its epicycle with respect to the celestial sphere in 8 years, relative to the stationary earth. Therefore, we can divide the number of days in 8 years by 13 circles to get the time for Venus to turn around its epicycle, that is, 225 days/circle.

By the same token, Mercury is in superior conjunction (or inferior conjunction) with the sun for 145 times in 46 years and 34 days, and Mercury has made 191 (= 46 + 145) circles around its epicycle in 46 years and 34 days relative to the stationary earth. So we can divide the number of days in 46 years and 34 days by 191 circles to work out the time for Mercury to make a circle around its epicycle, that is, 88 days/circle.

Now we have determined the speed of the epicycle and deferent of the five planets, and we still need to know the ratio of the radius between the epicycle and deferent of each planet. The planet's trajectory will not be retrograde if the ratio is too small, and the retrograde distance of the planet will be too long if the ratio is too big, so the appropriate ratio must be calculated according to observation to make sure that the retrograde distance of the planet consistent with the actual astronomical phenomena, which is tough work. After observation and calculation, Ptolemy calculated that the ratios of the epicycle-deferent radius of Saturn, Jupiter, Mars, Venus and Mercury are 6.5/60, 11.63/60, 39.5/60, 43.17/60 and 22.5/60 respectively. Please note that among the five ratios, the reciprocal of the first three ratios are 9.23, 5.16 and 1.52 respectively, and the last two ratios are 0.72 and 0.38, which are very close to the ratio of the five plants' orbit radius to that of the earth as we know now. This is absolutely not a coincidence. Remember what is mentioned before? There is another important method about the combination of the epicycle and deferent that has not been discussed, and we do not discuss it now. But we need to remember these five ratios, which are closely related to our later discussion.

With the accumulation of observation data, Ptolemy further discovered that there is obvious non-uniformity in the motion of the planet. Take Mars as an example. On some dates in a cycle of its deferent rotating around the earth, there is a distance difference of over twenty days between the actual observed position and the position it should reach on the epicycle, calculated according to the theory of uniform circular motion. Even the eccentric circle theory used by Hipparchus to explain the non-uniformity of the sun's motion does not apply to planets. Therefore, Ptolemy put forward the concept of equant. What is the equant? Let's take the amusement equipment as an example. In Fig. 16, two people are standing on both sides of the center column, one at point E and the other at point M. They are in a straight line with the center O of the center column, and the distance between them and the center

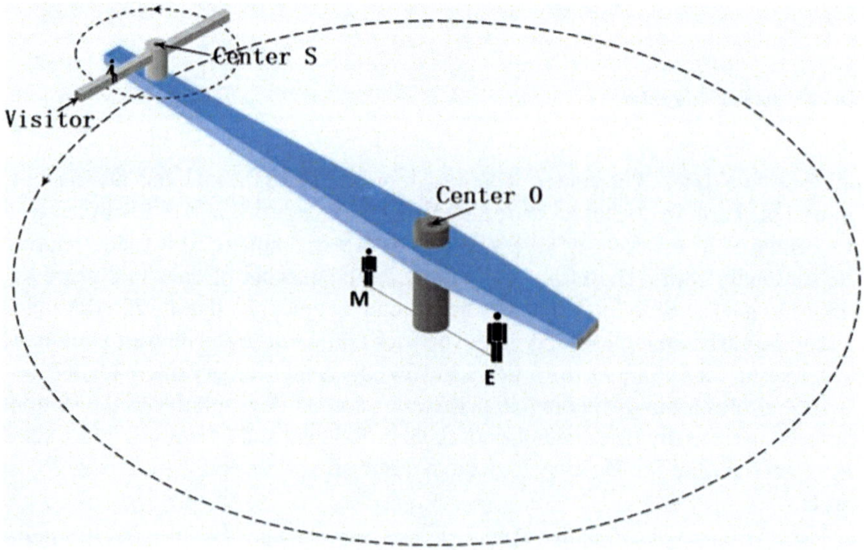

Fig. 16 Equant

O is equal. The deferent (long dashed line in Fig. 16) is an eccentric circle for the person standing at point E. In the previous discussion, the center S revolves uniformly around the center O, which means that the center of the epicycle moves uniformly on the deferent. Now we make some adjustments so that when the center S revolves around center O, it appears to be uniform for a person standing at point M, but the epicycle does not appear to move uniformly on deferent for the people who stand at the center of the circle (the center O) and point E. According to Ptolemy's model, tourists are planets, point E is equivalent to the position of the earth, and point M is the so-called equant. The point created by Ptolemy is indeed a little mysterious, but it has some connection with the two focuses of an ellipse, which shows that Ptolemy is not imagining without reason.

The combination of epicycle-deferent system, eccentric circle and equant is exactly Ptolemy's planet model.

In Ptolemy's later work called *Planetary Hypotheses*, he integrated his planetary models into a physically real system. He assumed that all possible heights in the sky were occupied by the planets. Each planet had its own height zone occupied by itself all the time, and these height zones neither overlap nor have gaps with each other.

Of course, there were many parameters to be considered by Ptolemy, such as the deviated distance between the earth and the center of the deferent of each planet, in which direction, and so on. Only when all the details are fully refined can this model be more practical. In a word, by carefully selecting the parameters of the big ball (deferent) and the small ball (epicycle), the model formed by the big ball and a small ball is consistent well with the actual observation at that time in terms of precision, and it is in accordance with Aristotle's system in physics. Therefore, Ptolemy's geocentric system can dominate the astronomical field for more than 1000 years.

Copernicus, a famous astronomer in the sixteenth century, commented, "Claudius Ptolemy of Alexandria, who far excels the rest by his wonderful skill and industry, brought this entire art almost to perfection with the help of observations extending over a period of more than four hundred years, so that there no longer seemed to be any gap which he had not closed."

Astronomer Michelle Hawkins, a professor at Cambridge University in England, commented, "Greek phase of the campaign to devise geometrical models that would replicate the motions of the seven planets. Future planetary positions could now be predicted encouragingly well, and there seemed no reason to doubt that adjustments made by later astronomers would bring prediction and observation ever closer".

Ptolemy absorbed and summarized the wisdom and achievements of ancient Greek astronomers, founding the geocentric system named after him, which was recognized by more and more astronomers and soon became the standard textbook for learning astronomy in the ancient western world. In the following 1300 years, no astronomer put forward any more reasonable and more convincing new universe model with the earth as its center.

The greatest success of this model is that it can accurately predict the position of celestial bodies at any time, which is definitely not a coincidence. Perhaps this is the model of the real universe? Or we are already close to the truth of the universe?

4 Questioning Ptolemy's Geocentric System

Ptolemy's geocentric system has been so precise and practical, but will other astronomers still have any doubts about it? In fact, the doubts about it have never stopped, and many problems exposed by it make it hard for other astronomers to believe that it is a model of the real universe. In other words, it still has a long way to go to discover the real universe.

1. Unrealistic Moon Model

The most obvious is Ptolemy's motion model of the moon. The epicycle, which carries the moon to move, is unusually large compared with the deferent which it moves on. In this way, the distance between the moon and earth can vary in the range as large as 33–64 times the radius of the earth, so the diameter of the moon should seem nearly doubled when it is closest to us compared with when it is farthest from us. As we all know the moon's apparent diameter changes very little. Ptolemy himself also admitted this flaw. Why didn't Ptolemy make the moon's epicycle smaller? Because it is not consistent with other celestial phenomena if the epicycle becomes smaller. He had no way to do it. It also shows that Ptolemy at least did not guess the true motion of the moon.

2. Lack of Harmony

Most astronomers have recognized the idea of harmony in the universe since Pythagoras put forward it. The celestial sphere revolves around the earth from east to west, while the sun, the moon and the five planets are not only "forced" to follow the celestial sphere to move with it, but also move backward from west to east against it. These two opposite motions are not harmonious anyway, and it has been an old problem, but Ptolemy cannot solve it either. Aristarchus argued eloquently in the third century BC that the volume of the sun is hundreds of times larger than that of the earth (see Chap. 5 for details). Ptolemy also calculated that the distance between the sun and the earth is 1210 times the earth's radius, and the radius of the sun is 5.5 times that of the earth, so the volume of the sun is about 170 times that of the earth. It is not controversial that the sun is much larger than the earth though the error of their calculation is too big. Is it reasonable for such a large celestial body to revolve around a small celestial body?

3. Unreasonable Order of Celestial Bodies

From Plato to Aristotle to Ptolemy, the order of the five planets in their universe model is a noticeable and interesting issue. Plato believed that the order of celestial bodies from the earth is as follows: moon, sun, Mercury, Venus, Mars, Jupiter, Saturn and stars. Aristotle believed that the order should be as follows: moon, Mercury, Venus, sun, Mars, Saturn and stars, and so did Ptolemy. The controversy lies in Venus and Mercury, which are always around the sun, so the ancients thought that it also takes one year for them to make a circle around the sun. Plato thought that they are above the sun and below Mars, while Aristotle thought that they are below the sun and the moon. Some astronomers thought Venus is above the sun and Mercury is below the sun. Opinions varied a lot, but none of them could explain why. Ptolemy thought that Venus is below the sun and Mercury is below Venus, and he explained that the sun should move between planets that are in opposition with it and those not in opposition with it. This argument is unconvincing, and the contradiction lies in front of us, because the fact that the moon will be in opposition with the sun every month (that is, the time when the moon and the sun are 180° apart) reveals the fallacy of the above opinions. As for why Venus is up and Mercury is down, Ptolemy could not give any more reasons. What is wrong if Mercury is up and Venus is down? As Michelle Hawkins, author of the *Cambridge Illustrated History of Astronomy*, put it, "The order he selected for these two, putting Mercury below Venus, was based on little more than the toss of a coin." How can the random order of celestial bodies be believed?

4. Lack of Unity

As is mentioned before, planets are divided into two types in Ptolemy's universe model. The centers of the epicycle of Mercury and Venus are firmly tied to the earth-sun line, and it takes a whole year for them to make a circle along deferent. But the angular distance between Mars, Jupiter and Saturn and the sun can be random, and it takes a whole year for them to make a circle on their respective epicycle. The motions of the five planets all have the same annul factors as the sun, but two are on the deferent and the other three are on the epicycle, which is exactly opposite. Why are there two types? Why are the centers of the epicycle of Mercury and Venus firmly tied to the earth-sun line? Why are the lines between Mars, Jupiter and Saturn and centers of their epicycle always parallel to the earth-sun line? Why does the epicycle-deferent model of the five planets all contain the annual factor? Ptolemy could not tell the reasons. He described the apparent motion of each planet with a graphic method. In order to make the calculation more consistent with the actual observation results of celestial phenomena, he needed to deal with the planets one by one separately. There was a set of models composed of the sun, the moon, Mercury and Venus, and another set of models composed of Mars, Jupiter and Saturn, so it is confusing as a complete system. Therefore, Copernicus said, "Their experience was just like someone taking from various places hands, feet, a head, and other pieces, very well depicted, it may be, but not for the representation of a single person; since these fragments would not belong to one another at all, a monster rather than a man would be put together from them."

5. Mysterious Epicycle and Deferent

Ptolemy's geocentric system could not determine the absolute size of the epicycle and deferent, and he just needed to know the ratio of the radius of each epicycle to deferent. In other words, the deferent of Saturn can be drawn larger or smaller on Ptolemy's model, and it will not affect the calculation and prediction of Saturn's position, as long as the ratio of epicycle to deferent remains unchanged. We have discussed this feature of the epicycle-deferent system in Chap. 5. Please remember it and we will discuss it seriously later. It is the same with other planets, and it does not matter whether you draw them big or small. You can even draw Jupiter outside

Saturn, Mars inside Venus, or disturb the original order, and it will not affect the calculation and prediction. In a word, you can draw them as you like, because all kinds of celestial phenomena do not depend on the size and order of the planets' orbits. However, the size of the orbit must be specific in the real physical world, but Ptolemy did not know it and could not calculate it. They are forever a mystery.

6. Imaginary Geometric Point

Ptolemy created the equant to solve the derivation caused by the uniform motions of planets, which could really solve big problems, but many astronomers doubt if there is such a point in the real physical world? In the eleventh century, after the astronomers of the western Arabian School, the trend of thought against Ptolemy rose. It was initiated by Avempace, with Abi Bakr and Bitruji as its successors. They objected to Ptolemy's epicycle-deferent hypothesis on the ground that the planet must orbit a real material center, rather than a geometric point. It is almost impossible for such a huge celestial body to revolve around a simple geometric point in the real physical world.

7. Complexity

In Ptolemy's epicycle-deferent system, except for the sun, the moon and the five planets do not directly revolve around the earth, but around their respective epicycle whose center moves along deferent. This problem is not very complicated so far, but the earth is on some point that deviates from the center of deferent for a certain distance and in a certain direction, and it gets much more complicated when the epicycle-deferent system is combined with the eccentric circle. In addition, there is an inclination between the plane of each deferent and the plane of the ecliptic, and another inclination between the plane of each epicycle and deferent. The unequal values of the two inclinations as well as the introduction of equant make the problem very complicated. It is said that in the thirteenth century, Alfonso (1221–1284), a king of Castile of Spain, who was well versed in astronomy, felt that the structure of the Ptolemaic system was too complicated. He said that the order in heaven might have been better if God had asked him for advice when creating the world. He was thus accused of being a heretic and even removed from the throne.

The Ptolemaic system has not met real challenges for over one thousand years since its birth, and it has really been successful. However, its problems are also obvious, and the doubts on it have never stopped, and the questions above strongly affected the vitality of the Ptolemaic system. There is always questioning in the process of seeking scientific truth. The questioning is a kind of spirit and attitude, which also needs wisdom, and astronomy develops in the process of questioning step by step. Questioning the Ptolemaic system prompted later astronomers to repair it, perfect it, or completely change it.

With the Ptolemaic system as a backdrop, the new astronomical discoveries can deal a heavy blow to the geocentric system, as long as they can prove the mistake of the Ptolemaic system. The newly established system would be affirmed by other astronomers and philosophers provided that it could show more advantages in comparison with the Ptolemaic system.

Chapter 7
Put the Sun in the Center of the Universe

1 New System Arrived on the Scene

In those years, it was an important task for astronomers to answer the questions about Ptolemy's geocentric system and improve or fundamentally change it. We have learned that most astronomers made efforts to promote it, but no one has made fundamental changes to this system. In order to improve the accuracy of the planetary theory, astronomers constantly adjust the design of epicycle and deferent. Some of them added or subtract several small circles to the Ptolemaic system; some changed the revolving speed of the compound circle based on new measurements... However, Ptolemy's geocentric system has revealed so many problems that it was of no use to just repair it. But after over one thousand years, there was still little progress. Where is the breakthrough?

At that time, the Ptolemaic system has gone beyond the scope of science and passively put on the "armor" of religion. Shortly before Ptolemy founded his system, Christianity was born and started to develop in Europe. In the beginning, Christians rejected the theories of Aristotle and Ptolemy, because their viewpoints and attitudes of academic research were independent of theology, which disturbed the church deeply. In 1209 and 1215, the Christian Church, which had occupied the dominant position in Europe, twice decided to allow no one to copy, read or preserve the works of Aristotle and Ptolemy. But this high-pressure measurement failed to fully work, and new ideas were still spreading secretly. In 1231, the pope ordered to revise and annotate the philosophical and natural science works of ancient Greece, so Aristotle's philosophy was tampered into a theory to demonstrate the omnipotence of God, and Aristotle himself was praised as the ideological authority of the Christian world. Aristotle and Ptolemy's view that the earth is the center of the universe was favored by the church, and the Ptolemaic system was regarded as a unique truth to discuss the structure of the universe. Under this background, a fundamental change of the Ptolemaic system would not only be opposed by astronomers who supported it, but also by the church, which would require both sufficient reasons and extraordinary courage.

Y. Ge, *From Geocentric to Heliocentric: How Discoveries Are Made*,
https://doi.org/10.1007/978-981-16-7779-3_7

Who has such ability and courage? History calls for the emergence of great astronomers!

From the end of the 13th Century to the sixteenth century, the Renaissance in Europe greatly liberated people's minds and effectively promoted the development of literature, art and science. At the end of the fourteenth century and the beginning of the fifteenth century, capitalism sprouted in some cities along the Mediterranean coast. Successful businessmen and workshop owners believed more in personal value and power, society respected the spirit of innovation, enterprise and adventure, and knowledge and talents were highly valued by people. And the discovery of the New World in 1492 tremendously broadened the horizons of Europeans. In September 1519, Magellan's fleet sailed westward and returned to Spain in September 1522. Thus, the hypothesis that the earth is spherical has been proved most powerfully after two thousand years.

It was under this background that a new cosmic system was born. The astronomer who created this new system was Copernicus (Fig. 1).

Copernicus was born in Torun, Poland on February 19th, 1473. He lost his father in his early years and was raised by his uncle. It is said that Copernicus became interested in astronomy when he was in middle school, and he devoted great energy to astronomical research during his university and in the church later. After decades of research and thinking, he produced the scientific masterpiece *On the Revolution of the Heavenly Spheres* (*De revolutionibus orbium coelestium*). Fearing that he would be punished by religious forces for the publication of these new astronomical

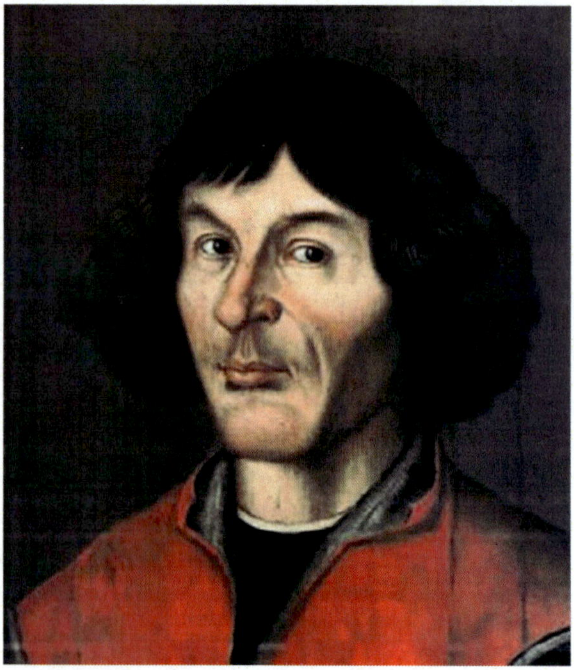

Fig. 1 Copernicus (the picture is from the Internet)

thoughts, Copernicus delayed publishing his works. After a long period of hesitation, Copernicus, with others' encouragement, finally published his *On the Revolution of the Heavenly Spheres* in Latin in Nuremberg, Germany, in 1543, just before his death. To relieve the pressure from the church, he presented this book to Pope Paul III.

Copernicus was not a professional astronomer, nor was he the astrologer that we mentioned earlier. He had his own career, so he was engaged in astronomical research in his spare time. He said that "among the many various literary and artistic pursuits which invigorate men's minds, the strongest affection and utmost zeal should, I think, promote the studies concerned with the most beautiful objects, most deserving to be known. This is the nature of the discipline which deals with the universe's divine revolutions, the asters' motions, sizes, distances, risings and settings, as well as the causes of the other phenomena in the sky". Astronomy was a lifelong endeavor for him.

Copernicus said, "I knew that others before me had been granted the freedom to imagine any circles whatever for the purpose of explaining the heavenly phenomena. Hence I thought that I too would be readily permitted to ascertain whether explanations sounder than those of my predecessors could be found for the revolution of the celestial spheres on the assumption of some motion of the earth."

Of course, no one has ever been to heaven, and all kinds of cosmic models created by predecessors are assumptions. Why can't he assume a new one? Moreover, there are so many problems in the previous assumptions, which indicates that those models are inconsistent with the real physical world. It is necessary to modify or make new assumptions. Maybe he can find more reliable assumptions. Indeed, anyone can assume freely, but the key lies in what your reason is and whether you can convince other people, otherwise your words will be regarded as nonsense by others. In particular, the assumption that the earth moves completely violates human intuition. Without sufficient reason, who will believe it?

What Copernicus called "some motion" is, first of all, the rotation of the earth, which is widely known today. This idea was not first created by Copernicus. Two thousand years ago, Hicetas and Ecphantus, two scholars of Pythagorean School, and later Heraclides put forward the hypothesis of earth's rotation, but it was opposed by most astronomers at that time. So why did Copernicus agree with their ideas?

Copernicus said, "If any motion is ascribed to the earth, in all things outside it the same motion will appear, but in the opposite direction, as though they were moving past it. Such in particular is the daily rotation, since it seems to involve the entire universe except the earth and what is around it." In other words, we feel that the whole universe is rotating around us while seeing the sun, the moon and stars rising in the east and setting in the west every day. It may be that the celestial body is really revolving around the earth, or the earth is rotating with us and we think it is the celestial body that revolves around the earth.

Why can't we feel ourselves moving at high speed at all? Copernicus said, "For when a ship is floating calmly along, the sailors see its motion mirrored in everything outside, while on the other hand they suppose that they are stationary, together with everything on board. In the same way, the motion of the earth can unquestionably produce the impression that the entire universe is rotating."

What he said should not be hard to understand. In ancient China, people also said that the boat goes without their awareness when people sit in a large ship with windows closed. We all have such experience when in a car, train, ship or plane. There were no cars, trains or planes in the age of Copernicus, but ships as the transportation vehicle had long been invented by human beings. Next, we have questions to discuss. Is the earth rotating or the celestial body revolving around the earth? What evidence does Copernicus have to prove that the earth rotates rather than celestial bodies revolve around the earth?

First, let's see the reasons against the rotation of the earth, which has lasted for nearly two thousand years. The earth is so huge that it is hard to imagine what it would be like if the earth rotates once a day according to people's established common sense. Imitating the tone of his opponents, Copernicus said, "A motion would have to be exceedingly violent and its speed unsurpassable to carry the entire circumference of the earth around in twenty-four hours. But things which undergo an abrupt rotation seem utterly unsuited to gather, and seem more likely, if they have been produced by combination, to fly apart unless they are held together by some bond. The earth would long ago have burst asunder, he says, and dropped out of the skies; and, what is more, living creatures and any other loose weights would by no means remain unshaken. Nor would objects falling in a straight line descend perpendicularly to their appointed place, which would meantime have been withdrawn by so rapid a movement. Moreover, clouds and anything else floating in the air would be seen drifting always westward."

It sounds really scary that the huge earth should rotate fast. According to calculation, the speed of people on the earth's equator will be 40 or 50 times faster than that of running horses. Then will we people and everything on the ground still exist? What a crazy idea it is! However, the huge earth is insignificant compared with the sky. When ancient Greek astronomers demonstrated that the earth is spherical, they also demonstrated that the horizontal circle divides the celestial sphere into two equal halves, which showed (see the proof in Chap. 4) that the sky is incomparably large and can be said to be infinite. Compared with the sky, the earth is only an extremely small point. If the earth is stationary and the celestial sphere is rotating, countless stars on the celestial sphere will circle the earth once a day at an unimaginable speed, which is faster than any object on the earth's surface, isn't it more incredible?

So, Copernicus asked, "But why does he not feel this apprehension even more for the universe, whose motion must be the swifter, the bigger the heavens are than the earth? … For the higher they are driven by the power of their motion, the faster that motion will be, since the circumference of which it must make the circuit in the period of twenty-four hours is constantly expanding; and, in turn, as the velocity of the motion mounts, the vastness of the heavens is enlarged. In this way the speed will increase the size, and the size the speed, to infinity." If so, the universe might have disintegrated long ago. Therefore, "Indeed, a rotation in twenty-four hours of the enormously vast universe should astonish us even more than a rotation of its least part, which is the earth." "It is more likely that the earth moves than that it is at rest".

Therefore, as long as you are willing to think about it, you will have to admit that it should be more reasonable to let the earth rotate. As Lomonosov, a famous Russian scholar and poet in the nineteenth century, said through the mouth of the cook in a limerick that he could also prove that Copernicus' words are right though he has not been to the sun. Who has seen the cook who let the stove turn around the barbecue?

Of course, the questions opposing the earth's rotation must also be answered. Why didn't the matter on the earth disintegrate because of the earth's rapid rotation? Why didn't the clouds and the things that hang in the air drift westward fast? Why didn't the arrow shot vertically fall far to the west? Copernicus answered these in his book, but to be fair, those answers could not be said to be very convincing, because the theory of physics had not yet developed adequately to answer such questions accurately at that time.

Then Copernicus wondered whether the annual motion of celestial bodies also can be attributed to another "some motion" of the earth, since the diurnal motion of celestial bodies can be attributed to the rotation of the earth.

Copernicus said that planets sometimes look bright and sometimes dark, indicating that they are sometimes close to and sometimes far from us. This fact firmly proves that the center of their orbits is not the center of the earth. It is obvious that Venus and Mercury revolve around the sun, and they cannot deviate from the sun more than their orbits allow, which can explain why the two planets always swing on both sides of the sun. The other three planets, including Saturn, Jupiter and Mars, are closest to the earth when they are in opposition with the sun (when the sun and the planet are on both sides of the earth), because the earth is between the planets and the sun at this time. On the contrary, these three planets are farthest to the earth when they seem to be near the sun (when the sun is between the planet and the earth). These facts suffice to show that these planets' center is the sun instead of the earth, which is consistent with the revolution of Venus and Mercury around the sun.

Seen from the earth, the apparent size of the sun does not change obviously, and its revolution around the earth is the most regular, which means that the distance between the earth and the sun has never changed greatly. There are only two possibilities. Either the sun revolves around the earth, or the earth revolves around the sun. Since all five planets revolve around the sun, it is more likely that the earth also revolves around the sun.

Therefore, Copernicus said, "Hence I feel no shame in asserting that", "Near the sun is the center of the universe", "The sun remains stationary, whatever appears as a motion of the sun is really due rather to the motion of the earth". "I suggest that we should now consider... it can be regarded as one of the planets." Hence in the Copernican system, the sun was placed at the center of the universe, and the earth was moved from the center to become a planet revolving around the sun, which is what Copernicus called the second "some motion" of the earth.

Regarding the earth as a planet was not first created by Copernicus. More than 1700 years ago, the ancient Greek astronomer Aristarchus had given the conjecture that the earth revolves around the sun. Due to various reasons, what we can know about Aristarchus's heliocentric conjecture is only a rough outline, while Copernicus' model of the universe centered on the sun is very complete. He described his model of the universe in poetic language.

"Of all things visible, the highest is the heaven of the fixed stars. This, I see, is doubted by nobody." "The first and the highest of all is the sphere of the fixed stars, which contains itself and everything, and is therefore immovable. It is unquestionably the place of the universe, to which the motion and position of all the other heavenly bodies are compared."

Below the sphere of the fixed stars is the first planet, that is, Saturn, followed by Jupiter, then Mars. The fourth position includes the earth and the moon as the epicycle, the fifth place is Venus, and the sixth position is occupied by Mercury.

"At rest, however, in the middle of everything is the sun. For in this most beautiful temple, who would place this lamp in another or better position than that from which it can light up the whole thing at the same time? For, the sun is not inappropriately called by some people the lantern of the universe, its mind by others, and its ruler by still others … Thus indeed, as though seated on a royal throne, the sun governs the family of planets revolving around it" (Fig. 2).

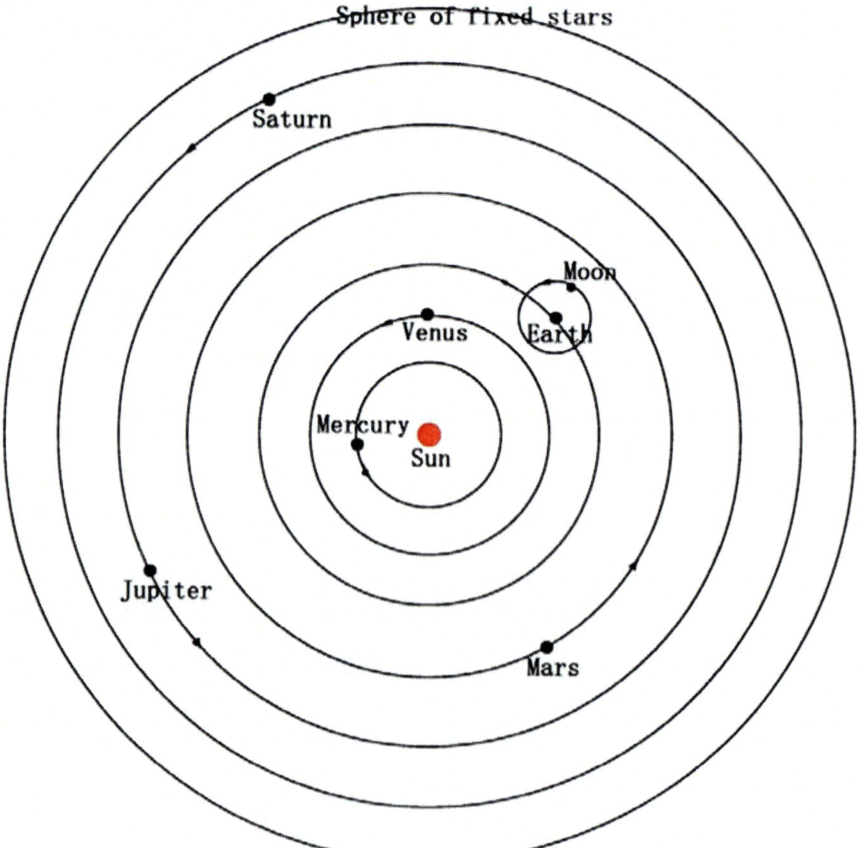

Fig. 2 Copernican heliocentrism model

In order to explain that all the parameters in Copernicus' universe model come from observation and calculation, and to prove his own universe system not only can explain all celestial phenomena, but also can predict the position of celestial bodies at any time, even better than Ptolemaic system, most chapters are "daunting" mathematical proofs in Copernicus' *On the Revolution of the Heavenly Spheres*. No wonder the publisher Johannes Petreius left a sentence in the preface, "Let no one untrained in geometry enter here". In addition, Copernicus also adopted the combination of epicycle and deferent to explain that planets are sometimes near and sometimes far from the sun. Unlike the Ptolemaic system, epicycles of planets in the Copernican system were very small (not drawn in the schematic diagram).

This is exactly what the Copernican system is: a completely new cosmic system, a system that challenges people's intuition, tradition and religion. Would anyone believe such a system? Would anyone dare to believe it?

2 Voice of Doubt

For any theory or viewpoint, people have every right to doubt it at first. It can be believed if it is persuasive, and it is a basic scientific attitude to doubt it and then believe it. Let's see whether Copernicus' reasons are tenable.

Copernicus' argument on the rotation of the earth is pretty convincing. Only by understanding that the earth is spherical can we conclude that the sky is bigger than the earth and it is incomparable. Only by understanding that the sky is bigger than the earth and it is incomparable can we realize that it is incredible for the nearly infinite sphere of fixed stars to revolve around the relatively tiny earth in 24 h. And we will admit that the earth is more likely to rotate than the movement of fixed stars.

No one has witnessed the earth rotating once a day around its own axis, and Copernicus also could not provide other direct evidence to prove that the earth is rotating. He could not even explain theoretically why we cannot feel the earth spinning at a high speed or why we have not been thrown away. However, he ruled out the possibility of a rotation in twenty-four hours of the enormously vast universe by strongly questioning, which left the earth with no choice but to rotate.

Compared with proving the rotation of the earth, Copernicus' argument on the revolution of the earth is much weaker, and in particular, he must answer an objection that lasted for more than 1700 years. If the earth revolves around the sun, it will move from one end of its orbit to the other every half a year, and the distance between the two ends is extremely far. When seen from either end, the relative positions of stars should change anyway, but why have astronomers never observed this change? (Fig. 3).

Copernicus' answer to this question is almost the same as that of Aristarchus, who lived more than 1700 years earlier, "Anyone who denies that the earth occupies the middle or center of the universe may nevertheless assert that its distance [therefrom] is insignificant in comparison with the distance of the sphere of the fixed stars." "This

Fig. 3 Stellar parallax

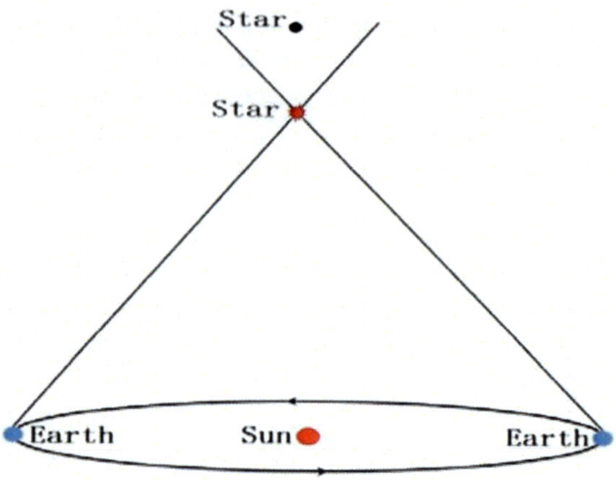

proves their immense height, which makes even the sphere of the annual motion, or its reflection, vanish from before our eyes."

What else can we say? The astronomical phenomena seen with the naked eye in the age of Copernicus are basically no different from those seen by Ptolemy 13 centuries ago and countless generations before Ptolemy. In these 1300 years, astronomers did not make any significant astronomical discoveries. Copernicus could not give direct evidence such as the annual parallax of stars to prove the earth's revolution, nor could he provide a physical basis for the earth's movement. It was really difficult to prove that the earth revolves around the sun.

It seems that, to put it bluntly, "the earth revolves around the sun" is merely Copernicus' "imaginative" and "reasonable" guess.

In Copernicus' heliocentric system, only the earth among the six planets has the moon revolving around it. Why is the earth so special? Does this mean that the earth is really different from other planets? Copernicus could not answer it, either.

The bigger problem is that, to deal with the non-uniformity of the speed of celestial body and the change of the distance between celestial bodies and the center of their orbit, Copernicus, who firmly believed that celestial bodies move uniformly in a circular orbit, had to retain Ptolemy's epicycle, and the parameters required by Copernican system are not much less than Ptolemy's. Copernicus had a large number of mathematical calculations in *On the Revolution of the Heavenly Spheres* to make a quantitative description of the movement of celestial bodies in line with his new system, which was as complex as the Ptolemaic system, and there was no improvement in accuracy. Since it is not better than the Ptolemaic system in use, what is its practical value? Why should people adopt this model?

3 Totally Enlightened by One Move

Theoretically, no one will believe it and will even oppose it when a new system has so many shortcomings, let alone facing great enemies who will attack it. But this is not the case at all. After Copernican system came into being, some people spent their whole life improving it, and some dared to risk persecution by the church to publicize it and provide new evidence for it. What do opponents and supporters see from this system?

1. Orderly System

More than 1700 years ago, Aristarchus eloquently demonstrated that the sun is much larger than the earth by using simple geometric knowledge, which has become an old problem that the geocentric system cannot avoid. It is difficult for people to believe that the sun, a relatively large celestial body, will revolve around the relatively small earth by simply relying on their experience and common sense. Copernican system has successfully solved this old problem. The huge sun has become the center of the universe, and the relatively small earth revolves around the relatively huge sun, which is obviously more reasonable.

The sun, moon and planets revolve from east to west with the celestial sphere, but they also move in the opposite direction from west to east. This is also an old problem that the stationary earth cannot overcome, which is obviously disharmonious. In the Copernican system, this problem does not exist any longer. All planets (including the earth) revolve around the sun in the same direction, which is unified and harmonious.

Copernicus put the sun at rest in the center of the universe and the earth moves. Once the earth that gets rid of the shackles of Ptolemaic system moves, the order of the universe will be fundamentally changed.

The order of celestial bodies has always been a difficult problem faced by astronomers. In the eyes of people on the earth, the five planets are just five small bright dots, and we cannot see which is near and which is far away no matter how hard we look at them.

(1) Sort by Motion Cycle

As is mentioned earlier, ancient astronomers arranged the order of celestial bodies according to their cycle of moving from west to east on the celestial sphere. Simple common-sense reasoning is that the farther the planet is from the center of the universe, the longer it takes to orbit the center. Saturn needs 30 years, so it is farthest from the earth. Jupiter needs 12 years, ranking second. Mars needs two years, and the sun needs one year while the moon needs one month. There is almost no debate about this order in history. At least the movement of the moon can directly support this view, because it has the shortest cycle and it will cover other celestial bodies during its motion, and its parallax is also the most obvious. All of these can prove that the moon is really the closest to us. But what about Venus and Mercury? They always move around the sun and can only be seen in the morning or at dusk. Most astronomers thought that their cycle of moving around the earth was the same as that of the sun, that is, one year. So, among Venus, Mercury and the sun, which is farther and which is closer?

Plato believed that Mercury and Venus should be above the sun and below Mars, while Aristotle and Ptolemy believed that Mercury and Venus should be below the sun and above the moon. However, none of them could give convincing reasons, especially their unreasonable words about the distance of Venus and Mercury to earth. For this reason, astronomers have been confused for nearly 2000 years.

In the Copernican system, the sun is at the center, and the five planets and the earth all revolve around the sun. Naturally, the order of celestial bodies should be arranged according to the cycle of them revolving around the sun.

Standing on the moving earth, how can we know the cycle of the five planets and the earth moving around the sun? Needless to say, the movement of the sun we see on the earth can be regarded as the reflection of the movement of the earth itself. In fact, the sun's revolution around the earth once a year reflects the earth's revolution around the sun once a year, which there should be no doubt about.

What about other planets? It would be much simpler if we could directly stand on the sun to observe the motion of the five planets, but it is impossible. Fortunately, this problem is not very difficult for Copernicus.

During the movement of Saturn, Jupiter and Mars, there will be a regular phenomenon of opposition, that is, the planets and the sun are on both sides of the earth. Seen from the earth, they are 180 degrees apart, which is called "opposition", and this phenomenon indicates that their orbit around the sun must be outside the earth's orbit. Whenever the opposition occurs, the planet, the earth and the sun are in a straight line. The planet's position on the celestial sphere seen on the earth is the same as that seen on the sun (see Fig. 4, taking Jupiter as an example). Collecting a certain amount of data about the position is equivalent to standing on the sun to have a long-term observation of planets, so it is not difficult to find out the cycle of their motion around the sun or even the speed of a certain period. There are many historical records of outer planets' opposition, hence we get the data that it takes 29.46 years for Saturn to revolve around the sun once, 11.86 years for Jupiter and 1.88 years for Mars, which is the same as the cycle around the earth of Ptolemaic system.

The cycle of Mercury and Venus's motion around the sun is also not hard to work out. They always move from side to side around the sun and will never be in opposition, which shows that their orbit around the sun must be inside the earth's orbit. Venus meets the earth 5 times in 8 years; in other words, Venus catches up with the earth for 5 times in 8 years. Venus actually makes 13 (8 + 5) circles around the sun in 8 years relative to the stationary sun, and divides the number of days of eight years by 13 circles, that is, 225 days/circle. Mercury meets the earth 145 times in 46 years and 34 days; in other words, Mercury catches up with the earth for 145 times in 46 years and 34 days. Mercury actually makes 191 (=46 + 145) circles around the sun in 46 years and 34 days relative to the stationary sun, and divides the number of days of 46 years and 34 days by 191 (=46 + 145) circles, that is, 88 days/circle. This is exactly the time the planets need to revolve around the so-called epicycle in the Ptolemaic system.

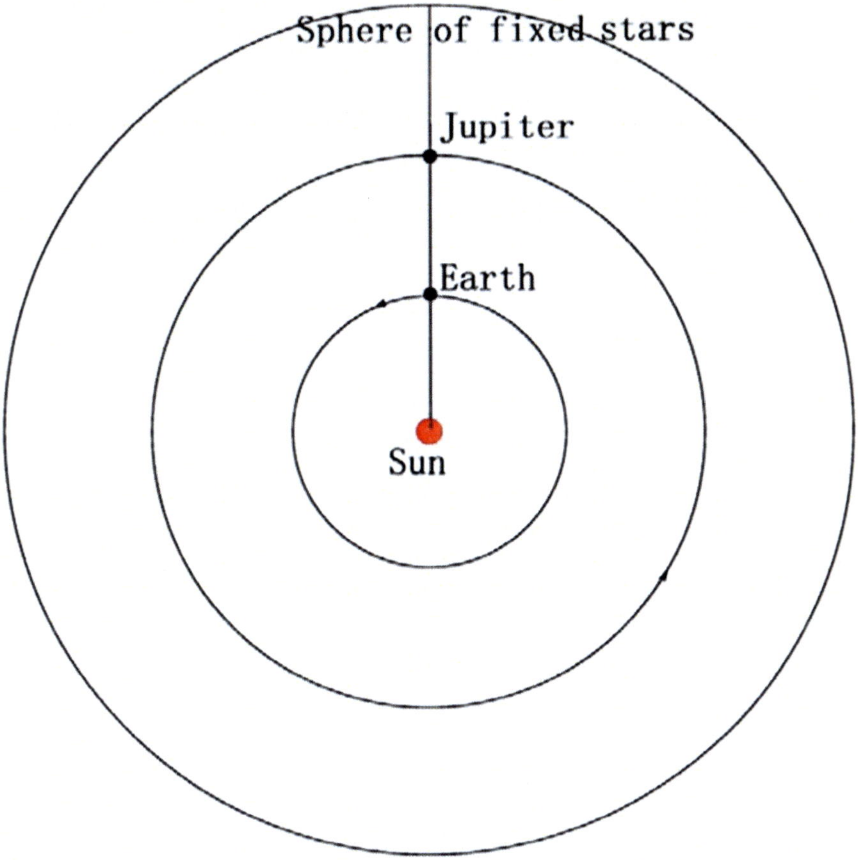

Fig. 4 Planet in opposition with the sun

Now that the cycle of six planets' motion around the sun is known, then the planets can be arranged according to their lengths of the cycle, which is as simple as the teacher queuing the students in class according to their height. The order of the planets is arranged according to their motion cycle from the sun as follows: Saturn (29.46 years), Jupiter (11.86 years), Mars (1.88 years), the earth and the moon (1 year), Venus (225 days), Mercury (88 days). The positions of Mercury and Venus have finally been determined after thousands of years, which can also well explain why the maximum angular distance of Venus from the sun is greater than that of Mercury from the sun, because the orbit of Venus is outside that of Mercury.

2. Sort by Size of Orbit

It seems to be a natural thing to sort according to the motion cycle, which has never been controversial, but objectively speaking, it is based on an inference that the farther away the planet is from the center, the larger its orbit is and the longer its motion cycle is. Must the truth be like this? It is hard to say. Why can't the planet close to the center revolve more slowly? The astronomer at that time could not explain it clearly. Therefore, this view can be doubted by others because it is somewhat arbitrary, although reasonable. Only when we know the true distance between the planets and the sun and arrange them according to the distance, that is, the size of the orbit, can we get a strict and reliable conclusion.

In the Ptolemaic system, the true size of epicycle and deferent remains a mystery all the time, and Ptolemy only needed to know the ratio of the radius of each epicycle to deferent, because the prediction of various astronomical phenomena did not depend on the size and order of planets' orbits. But Copernicus found a second criterion for the sorting of celestial bodies, that is, according to the size of planets' orbits. In the Copernican system, it is relatively easy to calculate the distance between each planet and the sun.

For Mercury and Venus, the relative distance (Fig. 5) between them and the sun can be roughly obtained as long as the maximum elongation between them and the sun is observed. Copernicus believed that both Mercury and Venus move in perfect round orbit. Therefore, when Venus reaches the maximum angular distance from the sun, the line between Venus and the earth must be tangent to the orbit circumference of Venus, and the line between Venus and the sun intersects at right angles to the line between Venus and the earth. Then, we can calculate the distance between point P and Point S in Fig. 5 if the degree of ∠E is observed in the right triangle SPE and we

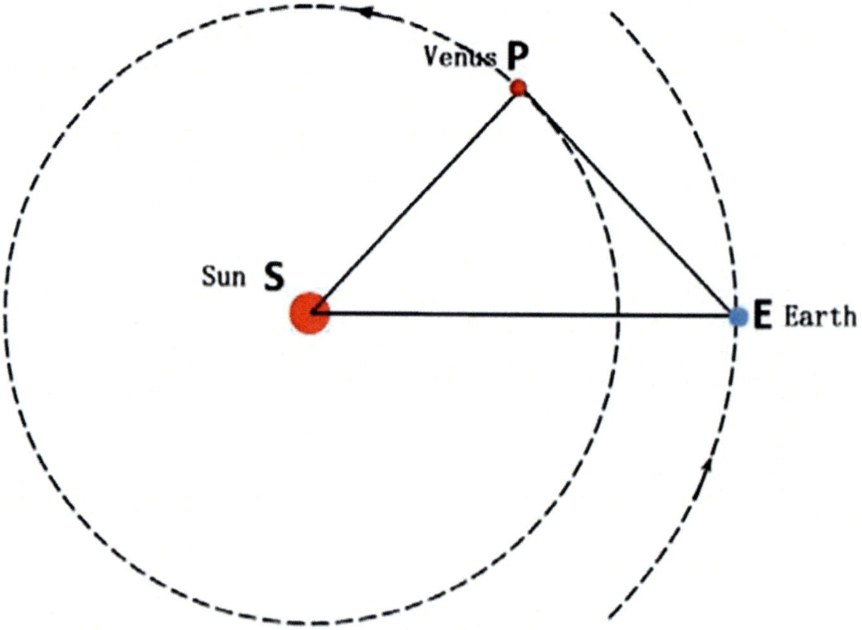

Fig. 5 Measurement of the distance between Venus and the sun

set the length of SE = 1 (please note that we will set the sun-earth distance to 1 in many later examples, because astronomers could not measure the absolute distance between the sun and the earth at that time, and they thus could only measure and calculate the relative distance).

In this way, Copernicus obtained that the distance between Venus and the sun is 0.72 times that between the sun and the earth, and the distance between Mercury and the sun is 0.38 times that between the sun and the earth.

What about the other three planets (Mars, Jupiter and Saturn)? Is there any way to measure their relative distance from the sun? This is an absolute problem in the Ptolemaic system, but it is not complicated for Copernicus. Copernicus said that their distance can be obtained through any parallax observation. Of course, it is better to do it when the planet is about 90° away from the sun, because at this time, the parallax of the planet is big and the calculation accuracy is high.

The parallax here refers to the difference between an outer planet seen from the sun and seen from the earth. When an outer planet is in opposition with the sun, the outer planet seen from the sun is consistent with that seen from the earth, and there is no parallax (Fig. 4). When an outer planet is in conjunction with the sun, this planet cannot be seen around the conjunction because the sun blocks the view from the earth. Except for these two cases, there is a difference between an outer planet seen from the sun and that seen from the earth at any other time, which is exactly parallax.

Copernicus listed three examples to calculate the distance between Saturn, Jupiter and Mars and the sun in *On the Revolution of Heavenly Spheres*. The examples are worth presenting because they highlight the superiority of this system and also show the wisdom of Copernicus, and the method of proving and calculating is not complex. Next, let's take Jupiter as an example to see how Copernicus calculated the data. Since the motion of the planets in Copernican system is still inseparable from the small epicycle, the calculation is somewhat complex, which has been appropriately simplified here (Fig. 6). From Fig. 6, we can directly see that the position of an outer planet on the sphere of fixed stars seen from the sun is different from that seen from the earth.

As is mentioned just now, it is not hard to work out the cycle of planets' motion around the sun or even the speed of a certain period by using the observation data of a certain number of the opposition between outer planets and the sun. Therefore, Copernicus could easily calculate the position of an outer planet on the sphere of fixed stars (seen from the sun) at a certain time, without really standing on the sun.

On February 19th, 1520, Copernicus carefully observed Jupiter six hours before noon. It was measured that the apparent position of Jupiter (point P) was 205° 9′ ecliptic longitude (vernal equinox was 0° ecliptic longitude), which was, seen from the earth (point E), at point D on the sphere of fixed stars. According to the law grasped of Jupiter's movement around the sun, it can be calculated that Jupiter (point P) should be on the point C of the sphere of fixed stars if we see it from the sun at this time, which is the true position of Jupiter in its orbit, that is, 194° 50′ ecliptic longitude.

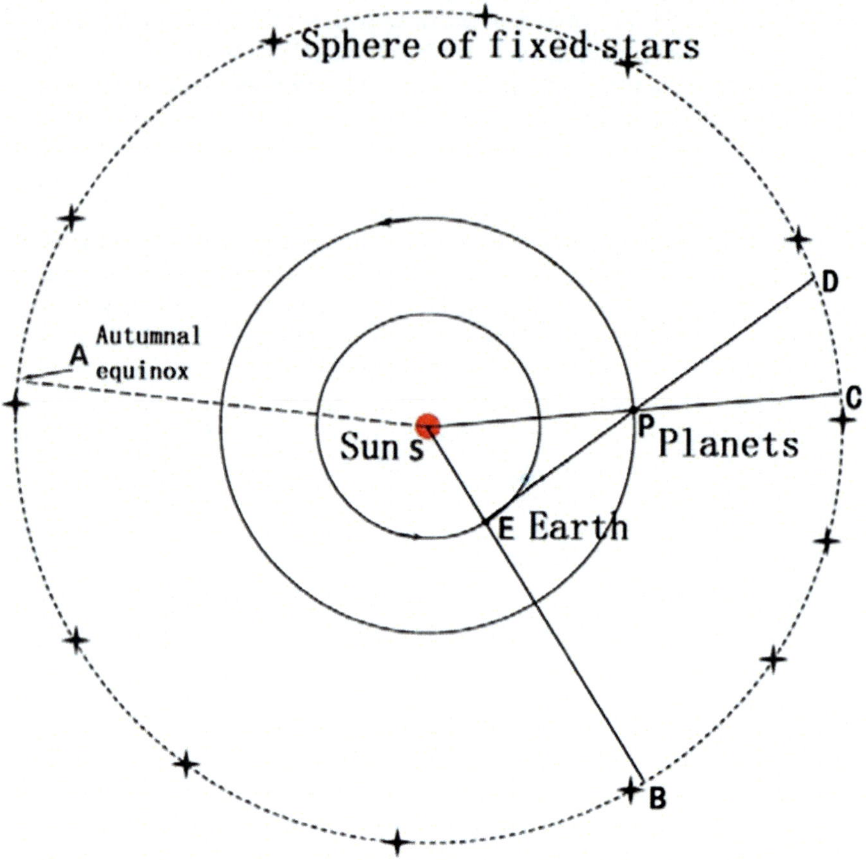

Fig. 6 Measurement of the distance between outer planets

Since the distance between point S and point C or D is almost infinitely greater than that between point S and point P, the angle between C and D seen from point P is almost the same as that seen from the center, that is, the sun (point S), so ∠DSC and ∠DPC can be regarded as equal, both equal to 10° 19′ (205° 9′–194° 50′).

Therefore, ∠SPE is 10° 19′, because ∠SPE is equal to ∠DPC, and this is the parallax of Jupiter at this time.

According to the law grasped of the sun's motion, we can calculate that at this time, the sun is 309° 16′ ecliptic longitude. Seen from the sun, the earth is at point B on the sphere of fixed stars, which is exactly 180° away from the position of the sun seen from the earth, namely 129° 16′ ecliptic longitude (309° 16′–180).

As a result, ∠CSB (∠PSE) equals 194° 50′–129° 16′, that is, 65° 34′.

∠SEP equals 180°–65° 34′–10° 19′, that is, 104° 7′; or 309° 16′–205° 9′, that is also 104° 7′.

Now, the degrees of the three angles in the triangle SEP are available.

Let's set the length of the sun-earth line (line SE in Fig. 6) as 1, and it can be calculated that the distance between Jupiter and the sun is 5.4 times the sun-earth distance according to the nature of triangle. This is the distance between Jupiter and the sun at this time, which is not exactly the same as the radius of Jupiter's orbit, because Jupiter's orbit is not a perfect circle. On this basis, Copernicus deduced that, when Jupiter is at the aphelion, the distance between Jupiter and the sun is 5.46 times the sun-earth distance and 4.98 times the sun-earth distance when Jupiter is at the perihelion.

Using the same method, Copernicus calculated that, when Saturn is at the aphelion, the distance between Saturn and the sun is 9.7 times the sun-earth distance, and 8.65 times the sun-earth distance when Saturn is at the perihelion. When Mars is at aphelion, the distance between Mars and the sun is 1.67 times the sun-earth distance, and 1.37 times the sun-earth distance when it is at the perihelion.

With the five planets' and the earth's distances to the sun (average radius of the orbit) obtained, it is very simple to sort the order of them. The outermost is Saturn (9.18 units on average), which is the farthest from the sun, followed by Jupiter (5.22 units on average), then Mars (1.52 units on average). The earth and the moon are the fourth farthest (1 unit), then Venus (0.72 units on average), and Mercury (0.38 units on average) is at the innermost, which is the closest to the sun. This is completely consistent with the original order according to the motion cycle, and there is finally a clearer standard for the position of the planets.

Copernicus said, "In this arrangement, therefore, we discover a marvelous symmetry of the universe, and an established harmonious linkage between the motion of the spheres and their size, such as can be found in no other way." "All these facts are disclosed to us by the principle governing the order in which the planets follow one another, and by the harmony of the entire universe, if only we look at the matter, as the saying goes, with both eyes."

This order is highly unified: the six planets, including the earth, follow the same law, and rotate around the sun in the same direction and along their respective orbits. They are not "forced" to move with the sphere of fixed stars, and there is no strange behavior of them moving forward and backward from time to time.

Such an order is so strict that "in no portion of it can anything be shifted without disrupting the remaining parts and the universe as a whole".

2. Explaining Celestial Phenomena in a Natural and Simple Way

Both the Ptolemaic system and the previous cosmic model must make reasonable explanations for various celestial phenomena, and the Copernican system is no exception. Can the Copernican system explain the various celestial phenomena we see? Of course! Not only so, it can also do it better and more naturally than all geocentric systems!

(1) Planets that Look Retrograde

The retrograde motion of planets has always been the primary problem perplexing astronomers, and all cosmic models must be tested by it. On this issue, Aristotle and Eudoxus's concentric sphere (also known as the crystal ball) model is better than Philolaus's central-fire system, and the epicycle-deferent system is better than the concentric sphere model. So, what better explanation can the Copernican system give? After all, there is no description of the retrograde motion of planets in Copernicus'model.

Copernicus'explanation is very simple. The six planets, including the earth, revolve around the sun, and the closer the planet is to the sun, the faster it moves, so the inner planets will surpass the outer planets again and again. And just in a short period of time around surpassing, if we stand on an inner planet to see an outer planet (or stand on an outer planet to see an inner planet), the motion of the outer planet (or the inner planet) is retrograde relative to the distant sphere of fixed stars (Fig. 7).

For us on the earth, Saturn, Jupiter and Mars always move backward when they are in opposition with the sun, and the position of a planet when in opposition is exactly the midpoint of its retrograde motion. Venus and Mercury always move backward when they are in inferior conjunction with the sun (the conjunction from evening star to morning star), and the position of a planet when in inferior conjunction is exactly the midpoint of its retrograde motion. This is in perfect agreement with our observations.

We will see that Jupiter moves backward longer than Saturn and shorter than Mars because it is closer to the earth than Saturn and farther from the earth than Mars. Similarly, since Venus is closer to the earth, we will see that it moves backward longer than Mercury on the sphere of fixed stars.

Therefore, the planets do not move forward or backward from time to time. The so-called mysterious retrograde motion of the planets is just because it looks so for us who stand on the moving earth.

In the Ptolemaic system, the planet is also at the midpoint of its retrograde motion when it is in opposition with the sun, but that is only because the rotational speed of its epicycle and deferent takes a specific value, so that the planet is in opposition with the sun when taken by epicycle to the closest position from the earth in the center. If the cycles of epicycle and deferent are somewhat different quantitatively, the qualitative law that makes the retrograde outer planet cross the sky and face the sun across the earth will not appear. But it will certainly appear in the Copernican system, regardless of the motion's specific rate at which the planet moves in orbit.

After thousands of years of exploration, human beings have been puzzled about the retrograde motion of the planet and spared no effort to work it out. Now they have finally got a very simple and natural answer in the Copernican system.

(2) Cause of Periodic Wobble of Planet

After long-term observation, astronomers knew the cycle of the planet moving along the zodiac in the background of the celestial sphere, but that is the average cycle. In actual observation, the interval between two adjacent returns of a planet to the same position on the ecliptic is not fixed. This is also a phenomenon that is difficult to understand and explain. Ptolemy dealt with this problem through geometric methods such as big epicycle and eccentric circle, which is very complex and cannot explain why.

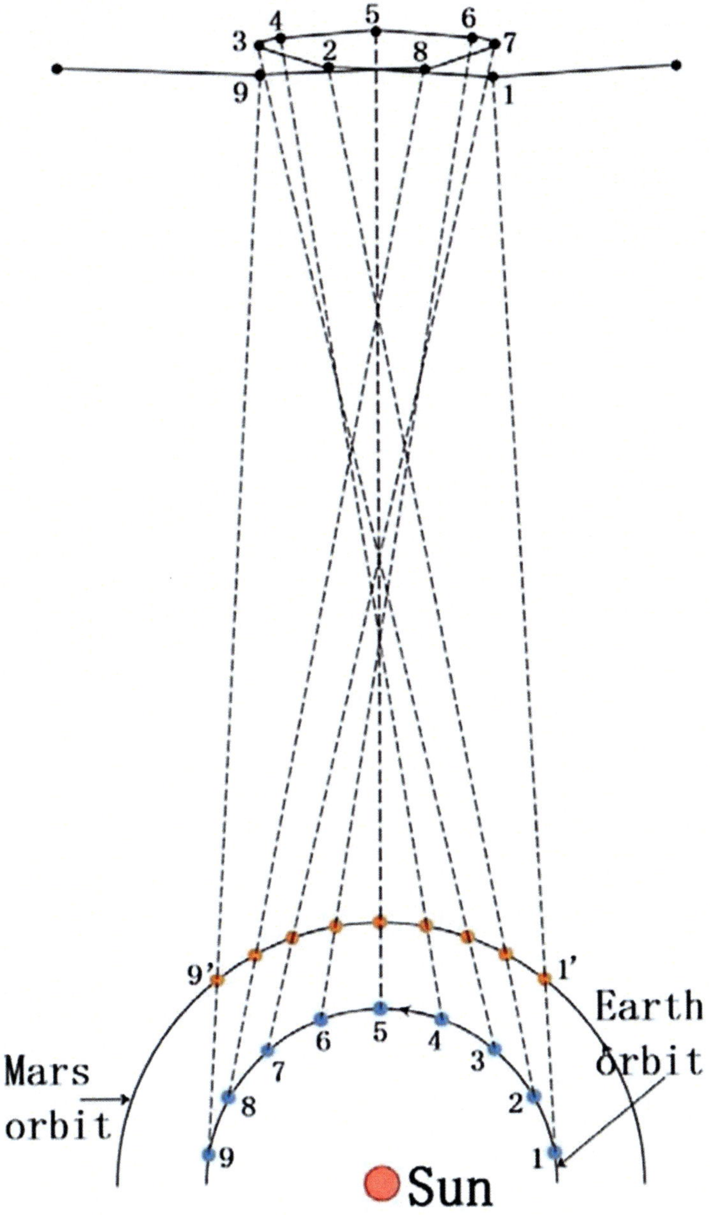

Fig. 7 An explanation of retrograde planets

The Copernican system answered this question well. We might as well use a simplified example to illustrate it.

Suppose that the motion cycle of an outer planet is 1.25 years, that is, five cycles of the earth's motion is exactly equal to four cycles of an outer planet's motion. Let's start to observe from the time when the outer planet is in opposition with the sun. Figure 8 is an explanation of the non-uniformity of the outer planet's cycle.

The outer planet is at point P and the earth is at point E_1. Seen from the earth, the outer planet is at point H_1 on the zodiac of the sphere of fixed stars. The outer planet returns to point P while the earth has made 1 and 1/4 circles to point E_2 after 1.25 years, and the outer planet is, seen from the earth, at point H_2 on the zodiac of

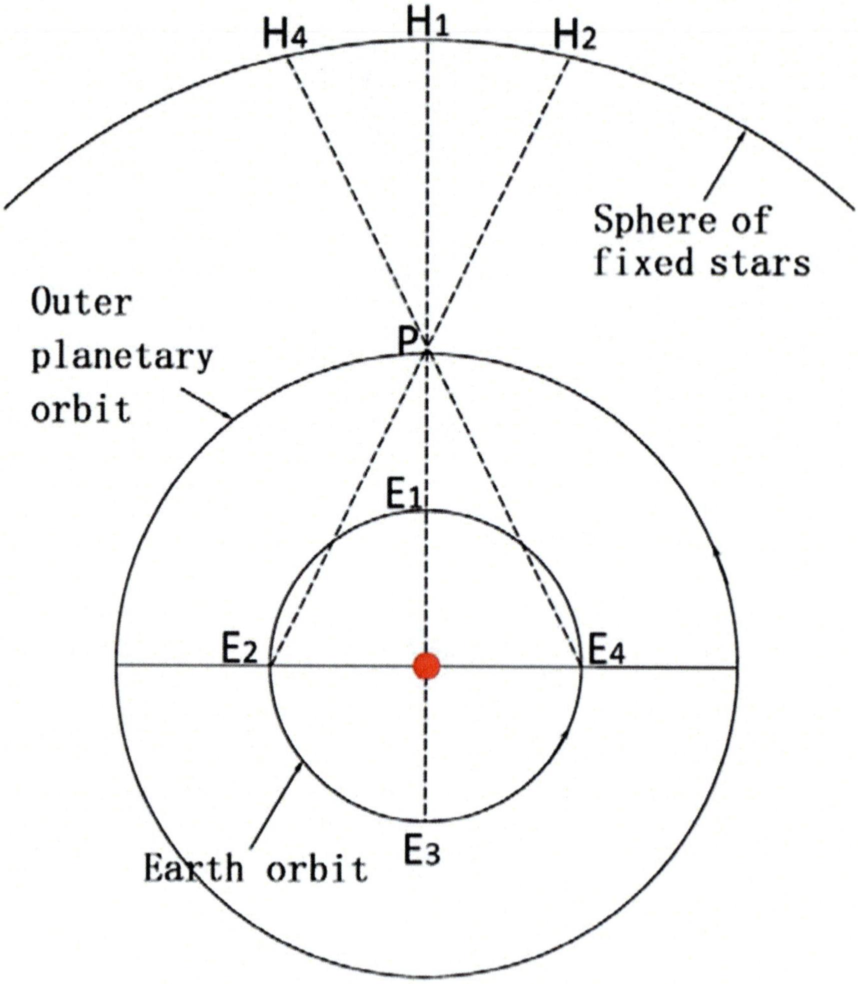

Fig. 8 An explanation of the non-uniformity of the outer planets' cycle

the sphere of fixed stars and there is still a short distance between the outer planet and point H_1. The outer planet returns to point P while the earth reaches point E_3 after another 1.25 years, and the outer planet, seen from the earth, is at point H_1 on the zodiac of the sphere of fixed stars. The outer planet returns to point P again while the earth reaches point E_4 after the next 1.25 years, and seen from the earth, the outer planet has travelled a slightly longer distance to point H_4 on the zodiac of the sphere of fixed stars. The outer planet returns to point P again while the earth exactly returns to point E_1 after the next 1.25 years, and seen from the earth, the outer planet is at point H_1 on the zodiac of the sphere of fixed stars.

In other words, seen from the moving earth, it takes more than 1.25 years for the outer planet to return to point H_1 for the first time, less than 1.25 years for the second and the third time, and more than 1.25 years for the fourth time. But in fact, the motion cycle of the outer planet is fixed.

Of course, the actual situation is much more complicated. None of the three outer planets' motion cycles around the sun are an integral multiple of that of the earth. But the reasoning is the same, which also applies to the inner planets. The interval between two adjacent returns of a planet to the same position on the ecliptic, seen from the earth, is not fixed, which is also because it just looks so for us on the moving earth.

Another complex problem has been explained simply and naturally.

(3) Why are Planets Divided into two Categories?

In the Ptolemaic system, the five planets are divided into two categories. Saturn, Jupiter and Mars need one year to make a circle around their respective epicycles, while the time for Venus and Mercury to make a circle around their respective epicycles is not one year, but it takes one year for the centers of their epicycles to make a circle around their respective deferents. Why are they divided into two opposite categories? Why does the motion mode of these two types of planets both contain the annual factor of the sun orbiting the earth? The Ptolemaic system cannot answer these. Moreover, it is an additional condition that the centers of Venus and Mercury's epicycles are fixed on the sun-earth line forever in the Ptolemaic system, and only in this way can Venus and Mercury be limited to move back and forth around the sun, which can be consistent with the observation. But why are only Venus and Mercury limited, while the other three planets are free? The Ptolemaic system cannot give a satisfactory answer.

All of these have become a question that can be simply explained in the Copernican system. Because the orbits of Saturn, Jupiter and Mars around the sun are outside the earth's orbit around the sun, and their angular distance from the sun is not limited, while the orbits of Venus and Mercury around the sun are inside the earth's orbit around the sun, and they can only move back and forth around the sun seen from the earth, without any conditions attached. Seen from the sun, the six planets (including the earth) revolve around the sun in the same direction in their respective orbits, and they belong to the same type. Since the earth revolves around the sun once a year, if we take the earth as a stationary reference while standing on the moving earth, the movements of the other five planets naturally have the factor of annual motion in our

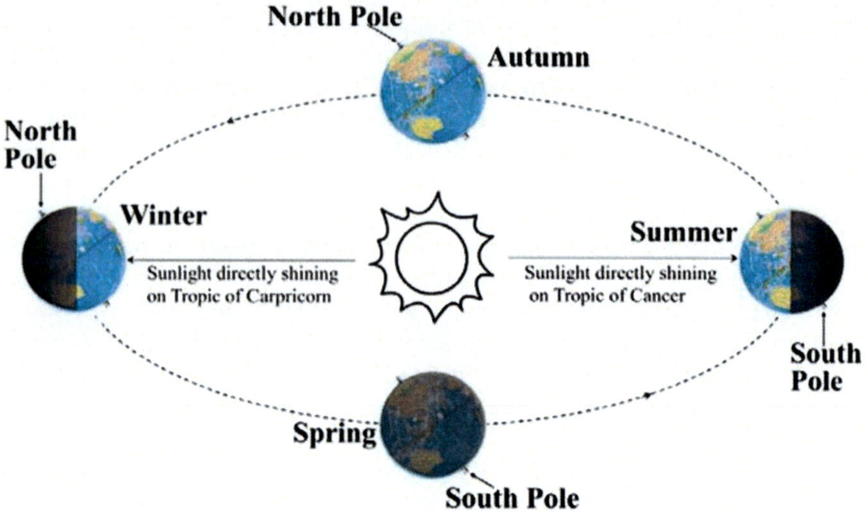

Fig. 9 Reasons for the changes of seasons

eyes. Among the five planets, one category of them moves outside the earth's orbit, which is called the outer planets; the other category moves inside the earth's orbit, which is called inner planets. The difference between the so-called two categories of planets lies only in this, nothing mysterious.

(4) Reasons for Changes of Seasons

Why are there four seasons in a year? Why do the rising point and setting point of the sun move slowly? Why is there a 20-degree angle between the ecliptic and the equator? The explanation given by the Copernican system for these problems is also very simple: such periodic changes occur because there is an included angle between the earth's rotation axis and the orbital plane of the earth's revolution. Figure 9 shows the reasons for the changes in seasons.

3. A Simple Model

Compared with the Ptolemaic system, the Copernican system is very simple because it has no Primum Mobile, no big epicycle, no conditions attached, and it does not need equant. Copernicus said, "This should be admitted, I believe, in preference to perplexing the mind with an almost infinite multitude of spheres, as must be done by those who kept the earth in the middle of the universe."

However, these words seem not fair. In the schematic diagram of the Ptolemaic system (Fig. 12 in Chap. 6), each planet has an epicycle in addition to the deferent (which can be regarded as orbit), while the planets have no other circles except their orbits around the sun in the schematic diagram of Copernican system (Fig. 2). But in fact, every planet in the Copernican system also has an epicycle, and it is unfair to hide the epicycle in the Copernican system when comparing it with the Ptolemaic system. Hence, many astronomers do not think that the Copernican system is simpler than Ptolemy's.

In fact, it is not difficult to see that this comparison is equitable as long as we analyze it carefully. Figure 10 is a comparative diagram of Jupiter's trajectories in the two systems. The big epicycle in the Ptolemaic system is mainly to describe the retrograde motion of the planet, and the combination of big epicycle and deferent makes the motion trajectory of the planet periodical (Jupiter as an example) as shown in Fig. 10a. The small epicycle in the Copernican system is to explain the non-uniformity of the planet's speed. The center of the small epicycle moves on the eccentric circle (counterclockwise in Fig. 10), and the planet moves on the small epicycle in the same direction (counterclockwise). The speed of the epicycle is twice that of the deferent, that is, the epicycle makes two circles when the deferent makes one circle. The uniform motion of the epicycle on the eccentric circle and the uniform motion of the planet on the epicycle make the planet's motion trajectory an almost complete circle, which has been discussed while explaining the features of epicycle and deferent (Fig. 6a in Chap. 6). We can see from Fig. 10b that the black spot is the track of the planet's motion, which forms an almost complete circle, but the upper part is dense in the term of the density of black spots. Since the moving time of the planet between two adjacent black spots in Fig. 10 is equal, the dense black spots indicate that the planet moves more slowly while the sparse lower part indicates that the planet moves faster. The sun is below the center of the circumference, which is consistent with the observation, that is, the planet moves faster near the sun and more slowly far from the sun. Comparing the two diagrams, we can see that the Copernican system is obviously simpler than the Ptolemaic system in terms of the track of the planet's motion.

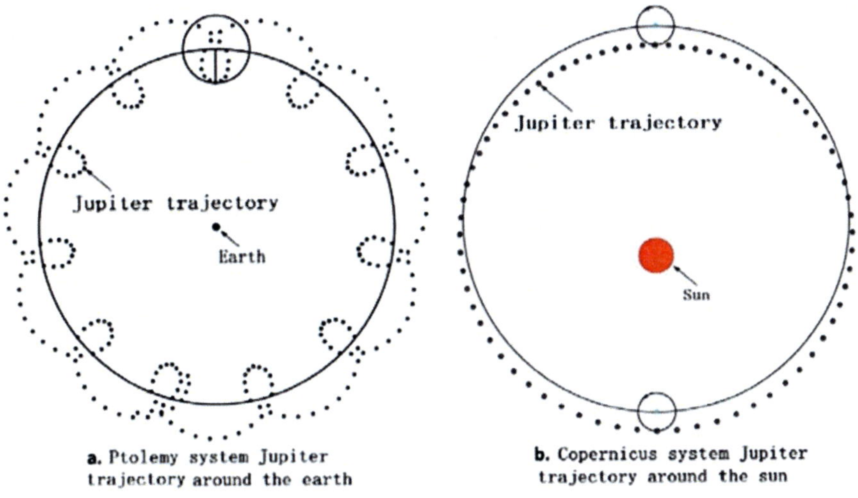

Fig. 10 A comparative diagram of Jupiter's trajectories in the two systems

The Ptolemaic system has finally encountered a real challenge after more than 1300 years since its establishment!

After learning the Copernican system and his reasons, do you think his assumption is more reliable? Some scholars say that Copernicus' new planetary system is a failure from a purely practical point of view. It is not more precise than that of predecessors who agree with Ptolemy, nor is it significantly simplified.

However, you will find the harmony, unity, naturalness and simplicity of the Copernican system when you look at it from another perspective. Many astronomers have believed that the universe is harmonious and simple since Pythagoras in ancient Greece, and the Copernican system shows a high degree of harmony, unity, naturalness and simplicity, which is unmatched by the Ptolemaic system. It is because of the remarkable harmony, unity, nature and simplicity that the Copernican system has gradually convinced more and more astronomers and philosophers. The vitality of the Copernican system is exactly reflected in these.

Chapter 8
Combination of the Two Major Systems

1 Eclectic Tychonic System

The publication of *On the Revolution of Heavenly Spheres* did not cause as much sensation as expected, but its impact was obvious. Fifty years after Copernicus' death, many advanced astronomy textbooks called him "the second Ptolemy" or "the great master of our times". These works increasingly borrowed and used data, calculations and charts from *On the Revolution of Heavenly Spheres*, at least from some parts that have nothing to do with the motion of the earth.

In 1551, only eight years after the publication of *On the Revolution of Heavenly Spheres*, Reinhold Erasmus calculated with the mathematical method established by Copernicus to release a whole set of new star catalogue, which soon became indispensable to astronomers and astrologers. More and more astronomers or people interested in astronomy started to read Copernicus' work, and gradually changed from believing in the Ptolemaic system to believing in the Copernican system.

Tycho Brahe (Fig. 1) was born in a noble family in Denmark in 1546, the third year after the publication of *On the Revolution of Heavenly Spheres*. At the age of 14, a solar eclipse aroused Tycho's interest in astronomy. In 1576, King Frederick of Denmark granted him jurisdiction of the island Ven, where he built the first major observatory in Christian Europe. During over 20 years on the island, Tycho developed instruments with increasing accuracy. Using these instruments, he and his assistants compiled unprecedented complete and accurate observation records that had never been seen before. The error of his measurement of the position of each planet is no more than $1'$, which has almost reached the limit that can be observed by the naked eye. He and the observers trained by him liberated European astronomy from its dependence on ancient data and eliminated a series of superficial astronomical problems caused by wrong data.

Tycho is also an astrologer. It is said that, when he was 20 years old, there happened to be a lunar eclipse (October, 28th, 1566) in Rostock, and after calculation, he declared that the lunar eclipse heralded the death of Suleiman, the Sultan of the Ottoman Turkish Empire. Soon, news of the sultan's death came. However, it was

Y. Ge, *From Geocentric to Heliocentric: How Discoveries Are Made*,
https://doi.org/10.1007/978-981-16-7779-3_8

Fig. 1 Tycho Brahe (The picture is from the Internet)

later learned that the 80-year-old sultan died before the lunar eclipse. Tycho made this prediction unknowingly in an era when communication methods and tools were very poor, which can also be regarded as "very accurate", right?

Tycho also needed to provide some services to the Danish royal family while he was engaged in his astronomical research on island Ven. For example, he made an astronomical report for Prince Christian. Tycho predicted that, the prince's childhood would be safe because Venus was in a favorable position. The prince would catch a minor illness in the second year after his birth due to Mercury's slightly poor position, but it would not be serious or dangerous. At the age of 12, the prince would catch a more serious disease caused by black bile. At the age of 29, the prince must be very careful in both health and dignity. The prince would face a great challenge when he was 56 years old, because neither the sun nor Mars had good intentions, and Venus could do nothing about it. If the prince could survive this disaster, he would have a happy old age.

However, Tycho emphasized at the end of each report that his prediction is not absolutely right, because God can change everything as he likes. We do not know whether he did this to free himself from punishment or to sincerely advise others not to believe in astrology too much.

Every astronomer is a potential cosmologist, and Tycho is no exception. He was familiar with the Copernican system and praised it as a "beautiful geometric structure". He admitted that it is easy to explain the motions of the five planets just by assuming that the earth moves. Copernicus frees us from the contradictions that mathematicians fell into in the past, and his theory can better explain the astronomical phenomena.

But Tycho refused to accept the view that the earth is moving, and he could not believe that the huge earth should move. Besides, he never found the parallax effect caused by the movement of the earth though he made a very careful observation. This means two possibilities, either the earth is still, or, as Copernicus said, the distance between stars is almost infinite. He did not believe that there would be such a vast empty space between the planetary system and the sphere of fixed stars, so he did not believe that the earth would move.

So how to keep the advantages of the Copernican system and avoid falling into the ridiculous situation of earth movement at the same time? In 1588, in a book about the great comet, Tycho "roughly" announced this system that later generations named after him (Fig. 2).

The earth is stationary, located in the center of the universe and surrounded by the moon and the sun. The other five planets, including Mercury, Venus, Mars, Jupiter and Saturn revolve around the sun in turn and around the earth driven by the sun. At the same time, the whole sphere of fixed stars, the sun, the five planets driven by it and the moon revolve around the earth day and night from east to west together. Of course, small epicycles and the eccentric circle are also necessary for the complete Tychonic system.

It is not difficult to see that this is an obvious mixed and eclectic system, which is not really an innovation. But it is also an almost perfect compromise. It adopts the quintessence of the Copernican system, making the five planets revolve around the sun, and maintaining the mathematical advantages of the Copernican System. And it retains the traditional view that the earth is still to meet people's intuition, avoiding the adverse factors in physics, cosmology and theology. The Tychonic system quickly replaced the Ptolemaic system because it maintained the geocentricism of the Ptolemaic system and was obviously superior to it. This is another geocentric system after Ptolemy. Unfortunately, it came too late, because the new Copernican system was trying to destroy the geocentric system.

2 Bridge Between Two Major Systems

Many people, including some great astronomers, paid little attention to the Tychonic system and thought it was just a compromise. However, if we analyze it carefully, we will be surprised to find that this seemingly eclectic system has built a bridge between

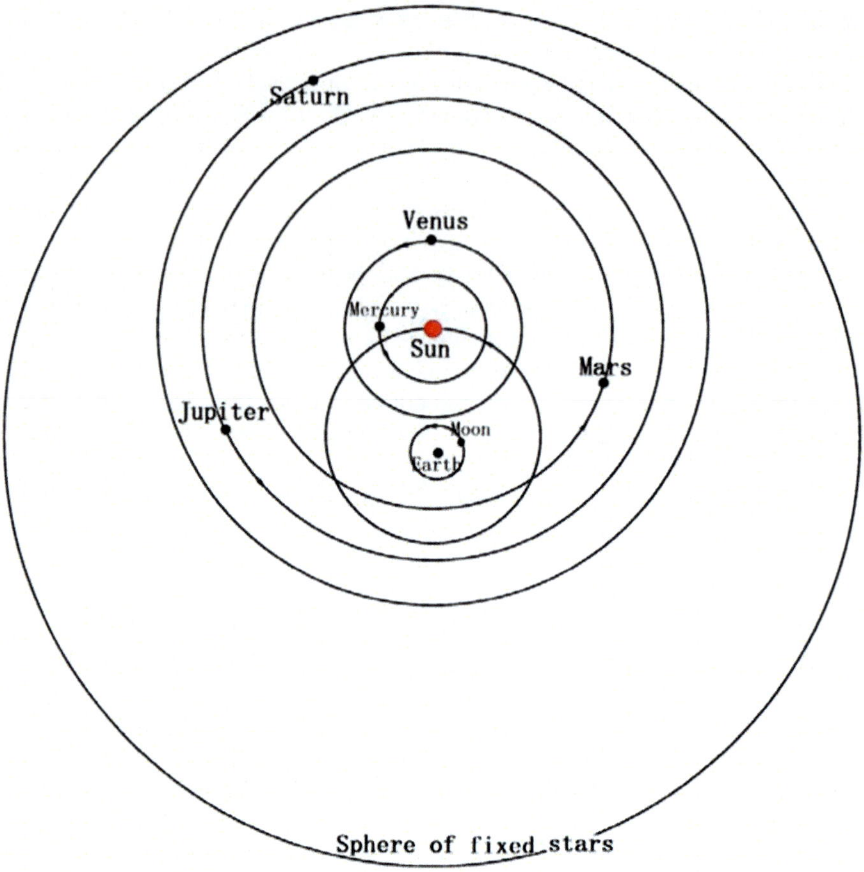

Fig. 2 Tychonic system

the Ptolemaic system and the Copernican system. There is actually an amazing internal relationship between the two systems that were opposing each other before.

As is mentioned earlier when discussing the epicycle-deferent system, there is still another important situation that has not been explained for the time being, and now it is time to do it. Let's take a look at Fig. 3. This is also large amusement equipment, which is completely the same as that we discussed earlier basically. The difference now is that the secondary rotating arm is obviously longer than the main rotating arm. The tourist sitting at the end of the second rotating arm makes a circle to draw an arc (long dashed line in Fig. 3), which covers the arc drawn by the rotation of center S around center O (short dashed line in Fig. 3), that is, the big epicycle covers the small deferent. In addition, the speed of the main rotating arm gets faster, while the speed of the secondary rotating arm gets slower and is slower than that of the main rotating arm.

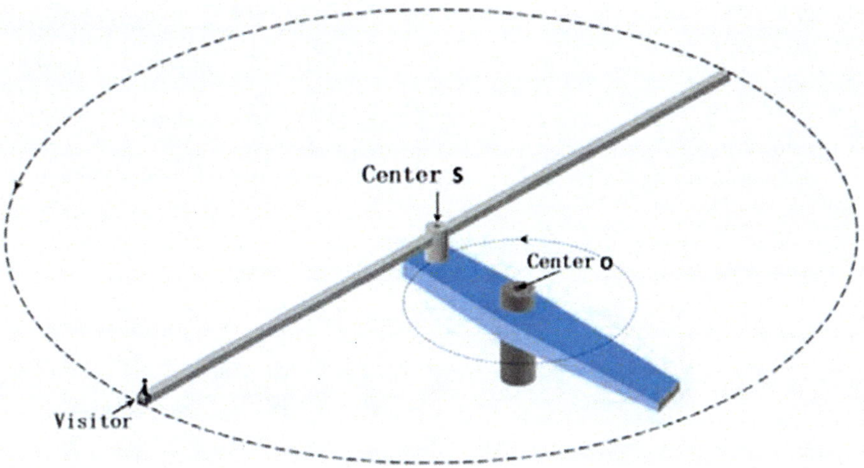

Fig. 3 The changed large-scale amusement equipment

What is the point of such a change? It can be clear if we choose two schemes to compare.

In the previous Chap. 6, we once assumed that the tourist sits at point B on the secondary rotating arm of the amusement equipment when we discussed the epicycle-deferent system. The distance from point B to point S is about 1/3 of the distance from center S to center O, and the main rotating arm makes a circle when the secondary rotating arm makes six circles. The track of the tourist's movement is a beautiful pattern formed by five petals (Fig. 7b in Chap. 6). Now we have modified the design to make the main rotating arm shorter and the secondary rotating arm longer, and the distance from center S to center O is about 1/3 of the distance from the tourist to center S; The revolving speeds of them also change, and the main rotating arm makes six circles when the second rotating arm makes a circle.

The radius and speed of the epicycle and deferent have been interchanged after such modification. What will the revolution track of the tourist be like now? We will be surprised to find that the tracks of the tourist in the two schemes are exactly the same, a pattern formed by five beautiful petals. Figure 4 is a simplified diagram of the two kinds of amusement equipment. Readers can compare their respective motion modes and the tourist's motion trajectory, or further move Fig. 4b to the left in imagination to overlap the two centers O, and you will see that a parallelogram is moving and changing. The vertex, that is, the center O is fixed, and the two adjacent vertices revolve around it, and the tourist revolves around the two vertices on the other vertex at the same time. From this, we can see that the two schemes depict the same trajectory.

This indicates that the epicycle and deferent can be exchanged with each other, which can also be proved mathematically. Since they can be exchanged with each other, can we try changing the Ptolemaic system? What will the changed Ptolemy's

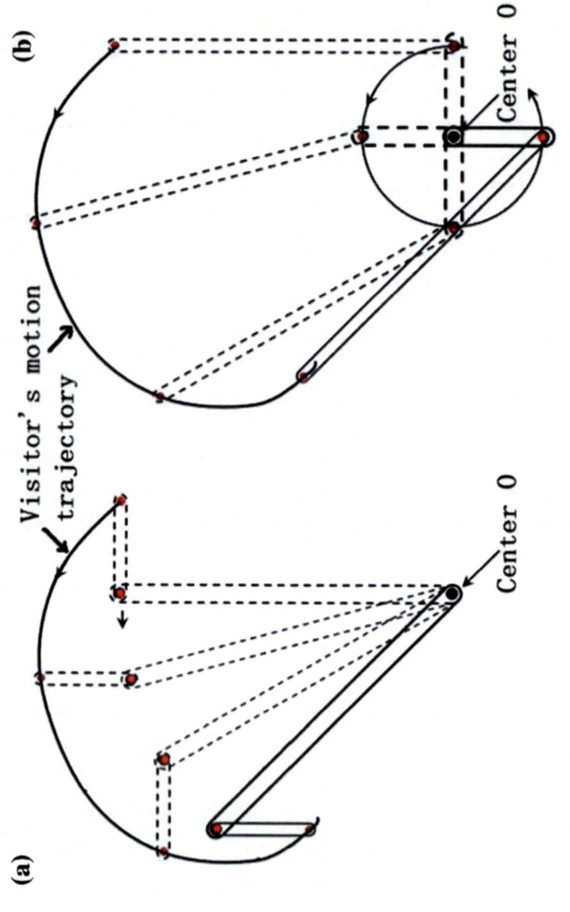

Fig. 4 Comparison of two amusement equipment's motions—depicting the same trajectory

system be like? As is mentioned earlier, Ptolemy only knew the ratio of epicycles to deferents of the five planets, and did not know their absolute size, but it did not matter and did not affect the calculation and prediction. We hence can be free to adjust the size of epicycle and deferent without changing the ratio.

First, we can set the radius of the sun's orbit around the earth as 1 unit, and then enlarge the deferent of Venus and Mercury to make them as large as the orbit of the sun. The radius of the epicycle must be a certain value if the radius of deferent is fixed, so how big should the epicycles of Venus and Mercury be? According to Ptolemy's calculation, the ratios of the radius of Venus and Mercury's epicycles to the radius of their deferent are 0.72 and 0.38 respectively, so the radius of Venus is 0.72 units and that of Mercury is 0.38 units. These two values are basically consistent with the relative values of the orbits of Venus and Mercury we know today or those calculated by Copernicus. The centers of the two planets were always on the sun-earth line before, now they have moved to the sun, and their speed remains unchanged. So, in fact, it takes 225 days for Venus to rotate around the sun and 88 days for Mercury to rotate around the sun. Mercury's orbit around the sun must be inside Venus's orbit around the sun because of the small epicycle of Mercury and the big epicycle of Venus. Such a change does not affect the use of the Ptolemaic system or the prediction of the position of Venus and Mercury.

Next, let's change the other three planets. First, we can adjust their epicycles to the same size as the sun's orbit around the earth. The radius of the deferent must be a certain value if the radius of the epicycle is fixed. According to the explanation of the Ptolemaic system earlier, Ptolemy calculated that the ratio of the radius of Saturn, Jupiter and Mars's epicycles to the radius of their deferent is 0.1083, 0.1938 and 0.6583 respectively. Now that the sizes of epicycles are set and the ratios are known, it is very simple to calculate the size of deferent. The radius of deferent of Saturn, Jupiter and Mars are 9.23, 5.16 and 1.52 respectively.

Now, the most crucial step is about to begin, which has never been paid attention to by any astronomer or mathematician in more than a thousand years from Ptolemy to Copernicus. Let's exchange the epicycles and deferents of these three planets so that their deferents are the same size as the sun's orbit around the earth, and their speed is the same as that of the sun, making a circle once a year. The radius of their epicycles are: 9.23 units for Saturn, 5.16 units for Jupiter and 1.52 units for Mars, and these values are consistent with the relative values of the orbits of the three planets we know today or those calculated by Copernicus. The epicycles of the three planets are larger than deferents and cover it, and their rotational speed on their respective epicycles are 29.5 years for Saturn, 11.86 years for Jupiter and 1.88 years for Mars.

In the Ptolemaic system, the lines between Saturn, Jupiter, Mars and the centers of their epicycles are always parallel to the sun-earth line. Now, it has to be changed that the lines between the centers of their epicycles and the centers of their deferents must always be parallel to the sun-earth line. There is only one possibility, that is, the overlap, which means that the epicycles' centers of the three planets must be on the sun, and their movements on their respective epicycles are actually equivalent to the revolution around the sun. According to the size of their epicycles or their motion cycle on their respective epicycles, Mars, Jupiter and Saturn are in order from the inside to the outside.

A miracle appears! The changed Ptolemaic system evolved into the Tychonic system. The five planets are arranged in orderly. The sun's orbit around the earth is their common deferents, and their respective epicycles are their orbits around the sun. They all revolve around the sun and follow the sun to revolve around the earth. Seen from the stationary earth, the orbits of the five planets are exactly a combination of epicycle and deferent.

With the same center of the earth and the same epicycle-deferent system, the Ptolemaic system has been transformed into the Tychonic system (Tycho slightly changed the ratio of epicycle to deferent according to his own accurate observation). Pierre-Simon Laplace, a French astronomer, said: "Such a simple and natural modification of the system of Ptolemy escaped all astronomers till the time of Copernicus. None of them appeared to have sufficiently considered the relations with subsist between the geocentric motion of the planets and that of the Sun, to have investigated its cause; none of them were curious to know their respective distance from the Sun and the Earth."

The relationship between the Tychonic system and the Copernican system becomes even simpler. Tycho took a step back from the Copernican system, and now we just need to take a step forward from the Tychonic system. But how to do it?

From the Tychonic system to the Copernican system, we only need to make the earth move. It is known that the motion is relative, so the observer must select a reference object while describing whether an object is moving, and then determine whether a certain object is moving according to the selected reference object. In other words, we can choose a coordinate system, and the choice of the coordinate system is only a method of description, which does not influence the natural phenomena to be described.

In order to describe the movement of celestial bodies, we can choose the earth as the center and put the origin of coordinates on it; we can also choose the sun as the center and put the origin of coordinates on it; we can even choose the moon as the center and put the origin of coordinates on it. It will be the Tychonic system if we put the origin of coordinates on the earth, and it will be the Copernican system if we put the origin of coordinates on the sun. The difference between the two systems is only an insignificant transformation of coordinate, and they are completely equivalent in mathematics. Of course, the earth in the Tychonic system must rotate to be truly equivalent to the Copernican system.

The Ptolemaic system can be changed into the Copernican system with only two times of transformations, isn't it amazing? For hundreds of years, many people once regarded the Ptolemaic system as "idealistic" or even "reactionary". Now it should be understood that it is a process and an "intermediate point" for scientists to seek the truth of the universe. Isn't the Copernican system so? The Copernican system of the universe we keep talking about is actually the structure of the solar system, which is like a grain of sand in the whole universe. Can we hence deny Copernicus' achievements?

The Tychonic system shows us the internal relationship between the Ptolemaic system and the Copernican system, and makes us know the scientific value of the Ptolemaic system and its status in the history of astronomy. The Tychonic system should have appeared after Ptolemy and before Copernicus, according to the order of ideological development. However, it is a pity that no one discovered such a relationship before Copernicus, neither Copernicus himself, nor Tycho, who put forward the Tychonic system, nor even Galileo.

3 Can History Be Assumed?

Let's make an interesting assumption. If an astronomer appeared in the long period of 1300 years between Ptolemy and Copernicus, and he found that "epicycle can be larger than deferent" and "epicycle and deferent can be exchanged with each other", which is not completely impossible, because it does not need other special historical conditions, as long as he can see and think of this. Therefore, the astronomer reasonably transformed the Ptolemaic system and put forward a new geocentric system such as the Tychonic system, which must be an important milestone in the history of astronomy. As a result, Copernicus, after this, no longer needed to take great risks to carry out a revolution, but only needed further conversion and improvement. If so, will the human understanding of the universe be more step-by-step and logical? If so, the Copernican system should not have encountered such strong opposition, and the development of astronomy would appear to be more moderate, wouldn't it?

We might as well follow this hypothesis to sort out, which may enable us to better understand the process of human understanding of the universe and see the footprints of exploration.

In the beginning, people thought that the earth was generally flat, but they could not imagine how deep the ground was, and could not explain how the sun, the moon and stars moved from west to east underground. The Pythagorean School of ancient Greece put forward the conjecture that the earth is spherical and demonstrated it, and they also pointed out that celestial bodies move at a uniform speed on the circumference. Although there were still many questions, most astronomers agreed with it because of the reliable evidence, and the spherical earth was also the best explanation for how the sun, the moon and stars moved through the earth. Astronomers began to make various conjectures about the motion of celestial bodies, and the early model of the universe was simple, that is, celestial bodies moved around the earth on a circle. However, the simple circular motion could not explain the mysterious whereabouts of the planets. Therefore, Plato pointed out that the speed and direction of the planet's motion can be explained by the combination of uniform circular motion, so Eudoxus designed a complex sphere combination centered on the earth. This combination, with rough and complex structure, can only explain the retrograde motion of the planet in general, but cannot explain the phenomenon that the planet is sometimes near the earth and sometimes far from the earth. Apollonius found another method of circumferential combination, namely, the epicycle-deferent system. This powerful system, in addition to the later method of the eccentric circle of Hipparchus, can well explain the direction, distance and speed of the planet. Ptolemy adopted the method of epicycle-deferent system and eccentric circle, refined and quantified it, and combined it into a complete system, to successfully predict the position of the planets. This system already implied the relative distance between the planets and the sun, and the speed at which the planets move around the earth.

Attention! Our hypothesis is about to begin!

Sometime after Ptolemy, an astronomer found that epicycle and deferent can be exchanged with each other when he was thinking about various problems of the Ptolemaic system, and it did not affect the prediction of the positions of the sun, the moon and planets. So, he exchanged the epicycles of Saturn, Jupiter and Mars with their deferents, and consequently, the cycles of the three planets are all one year, the same as those of Mercury and Mars. Then he adjusted the deferents of the five planets to the same size as the orbit of the sun, and the Ptolemaic system, in this way, was naturally transformed into a system that five planets revolve around the sun and follow it to revolve around the earth. Many problems and phenomena that are hard to explain in the Ptolemaic system have been well-solved. In the sixteenth century, Copernicus felt it was incredible that the earth was still, and the huge sphere of fixed stars and the sun, the moon and planets made a circle around the earth every day at an unimaginable high speed when he studied astronomy. And he also found that the earth (and the moon) always moves between the orbit of Venus around the sun and the orbit of Mars around the sun when the sun leads the five planets to revolve around the earth. If the sun is still in the center of the universe and the earth moves, the earth's orbit should be between Venus and Mars, which is also indicated by the cycle of the earth and the relative distance between the earth and the sun. Therefore, Copernicus fixed the sun in this system, changed the sun's orbit around the earth into the earth's orbit around the sun, making the earth revolve once a year and rotate once a day, and then the new Copernican system naturally comes into being. No one can object to it, nor can the church, because the deduction of mathematics is the most rigorous and irrefutable.

With a brief review and assumption, we can see more clearly how difficult and firm the pace of human exploration of the structure of the universe is.

Of course, history cannot be assumed. The real history did not develop on this smoother shortcut, and the Tychonic system came late. It did not come from the improvement of the Ptolemaic system, but from the retrogression of the Copernican system. Therefore, it did not last long although it was once very popular.

Despite the easy transformation of coordinates, in principle, the reference system can be selected arbitrarily when we study the motion of an object. But we often need to consider the convenience of the research when selecting the reference system to make the description of motion as simple as possible. The Tychonic system, seen just from the schematic diagram, is not as harmonious and perfect as the Copernican system. It will be a very complex road map if we depict the trajectory of the planets revolving around the sun, following it to revolve around the earth in the Tychonic system, which is far less simple than the Copernican system.

Thomas Kuhn, an American astronomer, said, "The Tychonic system has incongruities all its own: most of the planets are badly off center' the geometric center of the universe is no longer the center for most of the celestial motions; and it is difficult to imagine any physical mechanism that could produce planetary motions even approximately like Brahe's."

The Copernican system is not directly derived from the Ptolemaic system, and it is a new revolution. Because of this, it was strongly opposed by people and the church.

Through the Tychonic system, we can imagine how incredible it was to most people (including many astronomers) at that time that the earth is moving. From this, we can understand what courage Copernicus needed to make the earth move. At the same time, it also makes us sincerely admire Heraclitus, Ecphantus, and Aristarchus of ancient Greece more than 2000 years ago, for their rich knowledge, extraordinary wisdom and brave spirit.

Tycho was a lifelong opponent of Copernicus' theories, and his great reputation delayed astronomers from turning to new theories. However, at the same time, he created an astronomical system that quickly replaced the Ptolemaic system. Objectively speaking, the Tychonic system also played a role in publicizing the Copernican system more or less.

Tycho's historical contribution is a large number of accurate measurements of celestial bodies, and his greater contribution is to discover Kepler shortly before his death and give his rich observation data to this Sky Legislator. It is said that on the last night of Tycho, he was unconscious, saying to Kepler again and again in a weak voice like a poem that does not let him down and does not fail to live up to life.

Chapter 9
Towards Victory

The new cosmic system still has a long way to go to obtain a complete victory. As is mentioned earlier, the mathematical operations in the Copernican system are very complex, and the accuracy is no better than that of the Ptolemaic system. Since Copernicus firmly believed that the celestial body moves at a uniform speed on the circumference, he had to use the small epicycle, and even introduced the epicycle of the epicycle (or the second epicycle, that is, epicyclets). In terms of the geometric model, he used no fewer circles than Ptolemy. In the Copernican system, the sun is at the center of the universe, but it is only close to the center, neither the geometric center surrounded by planets nor the physical center. These all became the reasons for astronomers to oppose the Copernican system, and if these problems cannot be well-solved, the Copernican system is only an "assertion" and a "hypothesis".

In addition, there were objections from people out of common sense. Many people could not accept it and did not dare to believe that the earth inhabited by human beings would move at a high speed. They laughed at this crazy idea.

The greater opposition came from the church, because the Copernican system was a serious challenge to Catholic doctrine. The Bible says that God created heaven and earth. If the earth is an ordinary planet like Saturn, Jupiter, Mars, Venus and Mercury, how can heaven and earth be distinguished? And where are hell and heaven? The Bible says that the earth is equally firm and cannot be shaken, so who dares to put Copernicus' authority above the Holy Spirit? Copernicus' worries were becoming a reality. Scholars who defended and spread the Copernican system would be persecuted by the church at any time.

It is encouraging that great scientists appeared one after another in more than a century after Copernicus' death. They bravely defended, spread and improved the Copernican system, pushed it forward, and finally won a complete victory.

Y. Ge, *From Geocentric to Heliocentric: How Discoveries Are Made*,
https://doi.org/10.1007/978-981-16-7779-3_9

1 Sky Legislator—Kepler

In 1571, Kepler (Fig. 1) was born in Val City southern Germany. During his college years, he developed a keen interest in mathematics and astronomy, and he became an ardent supporter of the Copernican system under the influence of his teachers. He said that he proved it is true from the bottom of his heart, and he appreciated its beauty with incredible happiness.

In 1600, when Kepler was 29, at the invitation of Tycho, he came to Prague with his family, as Tycho's assistant to help sort out observation data and compile the new star catalogue. Tycho needed Kepler's mathematical ability to help him complete his astronomical theory work, and Kepler needed Tycho's rich and accurate observation data to carry out his own research due to his visual impairment caused by childhood illness. Hence, an opponent and an advocate of the Copernican system came together dramatically. Unfortunately, Tycho died of illness the next year (1601). This astronomer known as the "king of astrology" left Kepler with a large number of accurate observation data accumulated in his life. He warned Kepler many times before his death that they must respect the observation facts!

Fig. 1 Kepler (The picture is from the Internet)

Kepler was also a famous astrologer and a genius who was fascinated by mysticism. He was deeply involved in astrology activities closely related to the major political and military situation at that time, which led to his reputation as a great astrologer.

Famous Wallenstein was a German Czech aristocrat, Catholic and military commander of the Holy Roman Empire in the Thirty Years' War (1618–1648). In 1608, someone came to Kepler and asked him to arrange a fortune-telling for an aristocrat who "did not want to say his name" and predict this man's future. After calculation, Kepler made the following prediction.

He is melancholy and alert, loves alchemy, magic and divination, despises the commandments and customs of mankind and all religions, and doubts everything, whether it is what God does or what people do... He is cruel, arrogant, dissolute, greedy and fierce to his subordinates, cheating everywhere, and being changeable. He is silent from time to time, irritable and aggressive... In adulthood, most of his bad habits will be worn away, and his unusual qualities will develop into strong abilities to do things. A strong desire for fame, wealth, status and power can also be seen in him. Therefore, he will have many powerful, unfavorable, open and hidden enemies, but most of them will be no match for him... He will become a very superstitious person, and by this superstitious method, he can attract a large group of people around him and be elected by the mob as their leader.

The aristocrat who asked for astrological prediction anonymously was Wallenstein. 16 years later, Wallenstein asked Kepler to supplement the details of his future destiny, when he was about to become the commander-in-chief of the coalition, but Kepler refused to do so. Surprisingly, Kepler's calculation for the command-in-chief was stopped in 1634, and he was unwilling to continue the calculation. But in that year, after reaching the peak of his achievements, the commander was dismissed because of suspicion. He was assassinated and died on February 25th, 1634. It is said that when Wallenstein was assassinated, he did not escape or defend himself, but let a long sword pierce his heart.

Kepler devoted his whole life to the study of the real orbit of the planets and their mathematical laws. He believed that the assumptions that astronomers seek must not only accurately predict phenomena, but also look in line with the laws of nature and be tenable in physics. First, Copernicus said, "the uniform motions of the epicycle on the eccentric circle, and of the planet on the epicycle, cause the planet to describe not a perfect, but an almost perfect, circle." What kind of geometry is "an almost perfect circle"? In other words, what is the real orbit of the planet? How can we measure it? Second, the motions of planets in their orbits are not uniform. They move fast when they are close to the sun and slowly when they are far away from the sun. What mathematical laws do these motions follow? Third, the planets close to the sun have a short cycle while the planets far from the sun have a long cycle. What quantitative relationship does it have and what secret does it hide?

Kepler, who had the accurate observation data of Tycho, used his intelligence and wisdom to complete his historical mission after decades of unremitting efforts.

First, what is the real orbit of the planet?

It is unimaginably hard for us to obtain the real orbit of the planets by standing on the moving earth to observe the moving planets. Moreover, neither the earth nor the planets are simply moving uniformly in a circle. Perhaps only when we stand outside the sky and look back can we see the true orbits of the planets, but this is impossible because people cannot leave the earth. Kepler keenly realized that it is necessary to know the real motion of the earth before it is possible to determine the real motions of other planets.

How to determine the true orbit of the earth? Kepler assumed that if there is a fixed lamp not far from the earth's orbit, it is possible to determine the real orbit of our earth by using the sun and the lamp, and Kepler chose Mars as such a lamp. How is this possible? Isn't Mars also in motion? Is there any way to keep Mars still?

Mars and all the planets move in an enclosed orbit, and it will return to the same position in the sky regularly. Doesn't Mars seem fixed there at these specific moments? (This is also the only way. Mars's cycle of rotation around the sun is 686.98 days. If we want to use Jupiter as a lamp, we have to measure it every 12 years, let alone Saturn.) If we determine the position of the earth at several specific moments, the orbit of the earth can be known, can't it? Figure 2 is a schematic diagram for measuring the earth's orbit with Mars. Let's select a specific time, that is, when Mars is in opposition with the sun, the sun S, the earth E and Mars H are in a straight line, and seen from the earth E1, the position of Mars H in the sphere of fixed stars is at point A. After a Martian year, Mars returns to point H and the earth reaches point E_2. In the triangle SE_2H, $\angle SE_2H$ can be determined through the rich observation data left by Tycho, and $\angle E_2SH = 180° - \angle SE_2A$ (the sum of the internal angles of the parallel lines is equal to 180°. Kepler regards the straight line SA and E_2A as

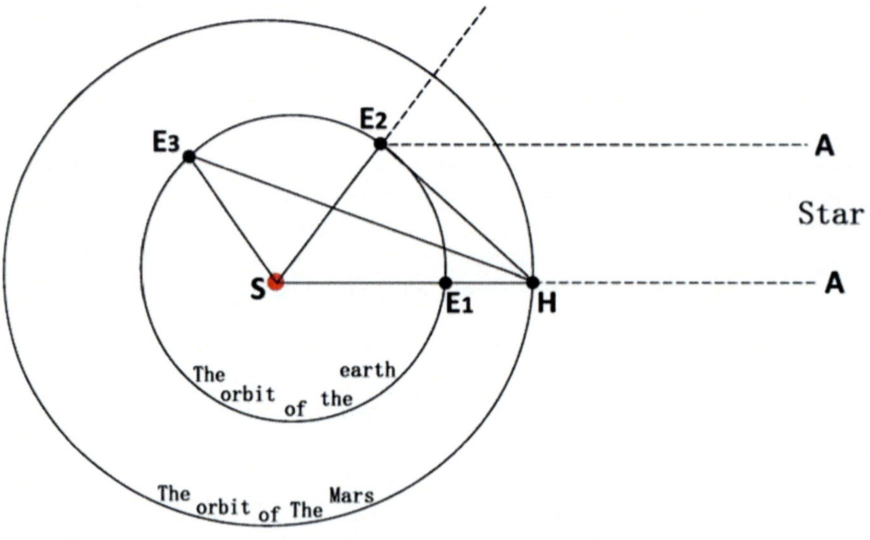

Fig. 2 Determination of the orbit of the earth by Mars

parallel because star A is very far away), and $\angle SE_2A$ can also be determined from the observation data of Tycho, so $\angle E_2SH$ can be calculated. In this way, the three angles of triangle SE_2H can be known. Let's set the length of SH to be 1 (unit), so the distance of SE_2 can be obtained, followed by the position of E_2. Another Martian year later, Mars returns to point H and the earth reaches point E_3, and the position of SE_3 can be obtained through the above method. Similarly, we can calculate the positions of E_4, E_5, E_6 and so on.

There were records of more than ten Martian years in Tycho's data, which meant that Kepler could get more than ten points of the earth's orbit through these data and calculations. What kind of figure would these points constitute? It was roughly like a circle, and Kepler did not doubt it. After checking, it was "indeed" a circle, but it was a little eccentric, which means that the sun was not at the center of the earth's circular orbit. Kepler needed to calculate where and how much it leaned. This was not very correct, but the good thing was that the earth's orbit was indeed very close to a circle, so "thinking" that it was a circle did not misdirect Kepler.

After obtaining the accurate orbit data of the earth, the position of the earth at any time and its distance from the sun has become known conditions, so it is easier to obtain the accurate positions of other planets at several time points and its accurate distance from the sun. We cannot help marveling at such a clever method!

Next, Kepler first studied Mars, because the orbit of this planet has always been the most difficult to grasp. He believed that he would be capable of understanding all planets if he could understand the orbit of Mars. Using Tycho's observation data, Kepler could calculate the position of Mars at several moments and its distance from the sun, but this is only a few points in the orbit of Mars. What kind of figure would these points constitute? It was also roughly like a circle, and Kepler did not doubt it at first. But some figures were inconsistent with the data in Tycho's observation data after many attempts and calculations. After about 70 times of calculations, he was almost close to the truth. Some calculated positions had an error of only 8' compared with Tycho's data. Could this small error be ignored? Or could this be a small error in Tycho's measurement? However, Kepler firmly believed that Tycho's data was accurate and there could not be such a big error, and this error could never be ignored. At last, he decided to abandon the assumption of circular orbit and to try to calculate the orbit of Mars with other geometric curves.

One experimental track after another was abandoned because it was found that they could not well match Tycho's observation data after complex calculations. Then, it was said that Kepler "accidentally" thought of an ellipse. What is an ellipse? Simply put, the trajectory of a moving point whose sum of the distance from two fixed points on the plane is equal to a constant is called an ellipse. Figure 3 is a schematic diagram of an elliptical orbit, in which Fig. 3a shows the method of making an elliptical orbit. It can be seen that the distance from any point on the ellipse to two fixed points is equal, and the two fixed points are the two foci of the ellipse. It can also be found from Fig. 3b, c that the naked eye cannot distinguish whether the real Martian orbit is a circle or an ellipse. We can hence know how difficult it is to calculate the orbit of Mars.

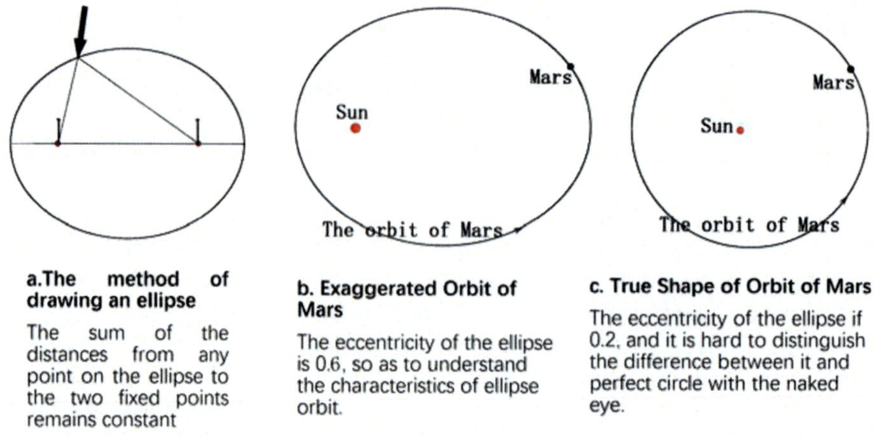

Fig. 3 Elliptical orbit

Finally, Kepler found that the orbit of Mars around the sun was not circular, but oval, and geometry helped astronomy a lot again. Then Kepler, after calculation, proved that the orbits of other planets are also ellipses. Among them, the eccentricity of Mercury's orbit is larger than that of Mars, and the eccentricity of other planets including earth's orbit is much smaller than that of Mars, which is more like a perfect circle. Therefore, Kepler might never find the true orbit of the planet if Tycho asked him to study the motion of Venus first.

Kepler broke the "spell" of circular motion that ruled astronomy for 2000 years. This is Kepler's first law of planetary motion, that is, the orbit of every planet is an ellipse with the Sun at one of the two foci.

Second, what is the law of planets' motions in orbit?

When Kepler still believed that the orbits of planets were eccentric circles, he was at the same time studying what mathematical laws govern the non-uniform motions of planets in their orbits. Both the earth and Mars move fast when they are close to the sun and slowly when they are far away from the sun. What is the relationship between speed and distance? After various assumptions and complex calculations, Kepler found such a law, that is, a line joining a planet and the Sun sweeps over equal areas during equal intervals of time. This is Kepler's second law of planetary motion, also the law that the area is proportional to time, and in fact, it came into being before the first law. Figure 4 shows Kepler's second law, in which the areas of the three shadows are equal.

The second law is a supplement to the first law. Without this supplement, given an orbit, it is almost or completely impossible to determine where the planet appears relative to stars at a specific time.

In 1609, Kepler published the book *Astronomia nova* and the article *On the Movement of Mars*, announcing these two famous laws. At the beginning of *Astronomia nova*, Kepler wrote, "Dear readers, please pity me if you are tired of this monotonous

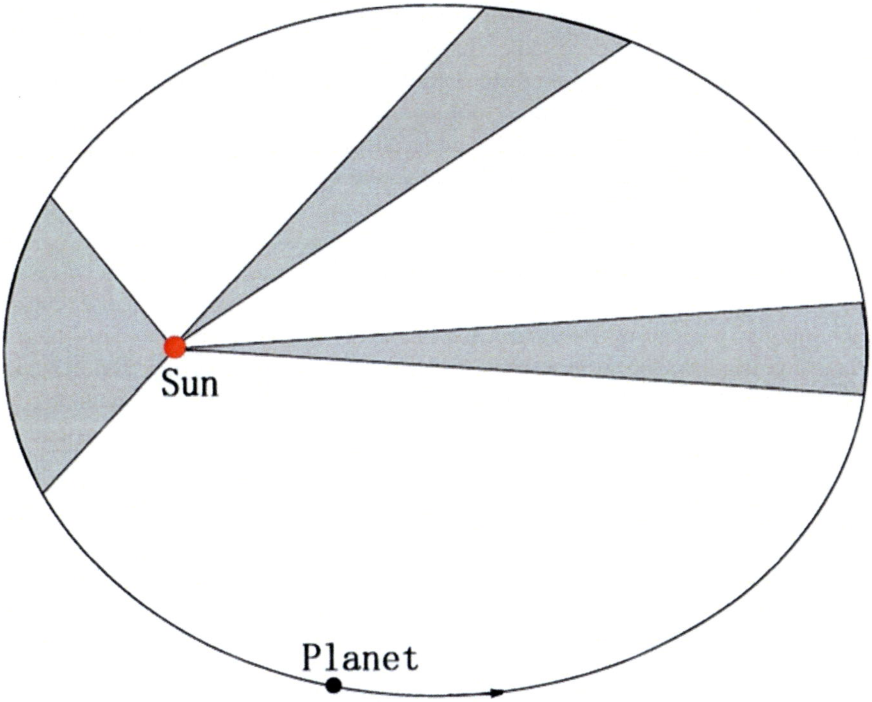

Fig. 4 Kepler's second law of planetary motion

process, because I have calculated at least 70 times." What hard work has Kepler done!

When the elliptical orbit replaces the basic circular orbit shared by the Ptolemaic system and Copernican system, and Kepler's second law replaces the uniform motion on the circumference, all these eccentric circles, epicycles, equants and other special devices are no longer needed. The Copernican system before was only concise in the qualitative description of the universe, and it was still indistinct in the quantitative description of details. Now, the Copernican system has really become concise and harmonious after Kepler's changes. Therefore, the Copernican astronomical system inherited by modern science is the product of the combination of Kepler and Copernicus.

The Copernican system achieved a key victory!

Third, what is the relationship between the distance of the planet from the sun and its cycle around the sun?

Kepler was most interested in this topic. When he was young, Kepler began to think about the universe created by God, and that Copernicus discovered the universe's layout of God, but failed to discover what prompted God to choose this layout over others. Why are there six planets instead of five or seven? What makes God arrange some planet to be a certain distance away from the sun rather than other distances? Why is there a certain speed rather than other speeds?

The initial research was interesting. Kepler thought that Euclidean geometry tells us that there are only five regular polyhedrons: regular tetrahedron, regular hexahedron, regular octahedron, regular dodecahedron and regular icosahedron. Each regular polyhedron can be surrounded by a sphere (circumscribed sphere), and each vertex of a regular polyhedron is on the circumference; Each regular polyhedron can surround a sphere (inscribed sphere), and each face of the regular polyhedron is tangent to the surface of the sphere. Will there be any relationship between only five regular polyhedrons and only six planets? Kepler believed that there must be a relationship because he believed that mathematics (especially geometry) was the language used by God. After careful study, Kepler arranged five regular polyhedrons and six planets as follows: Saturn's orbit is the outermost, inscribed with a regular hexahedron whose inscribed sphere is Jupiter's orbit. Jupiter's orbit is inscribed with a regular tetrahedron, whose inscribed sphere is the orbit of Mars. Mars's orbit is inscribed with a regular dodecahedron, whose inscribed sphere is the earth's orbit. The earth's orbit is inscribed with a regular icosahedron, whose inscribed sphere is the orbit of Venus. Venus's orbit is inscribed with a regular octahedron, whose inscribed sphere is Mercury's orbit.

The radius ratio of each spherical orbit obtained by this design is quite consistent with the known value of the size of each planet's orbit. Kepler was so happy with this wonderful model. In 1597, Kepler, at the age of 26, published this model in his book *Mysterium Cosmographicum*. But several years later, he found that the results calculated by this model could not be consistent with Tycho's data after he got the observation data of Tycho. Especially when he discovered that the real orbit of the planet was an ellipse, he completely gave up this wonderful model which had taken a lot of effort.

But Kepler firmly believed that there must be inherent laws that can be expressed mathematically in such an orderly arrangement of planetary orbits and cycles. He arranged each planet's orbital period and their average relative distance from the sun (Kepler only knew the relative distance, like Copernicus) into a table (Table 1), just like playing a numbers game. He added, subtracted, multiplied and divided these seemingly messy numbers and did all kinds of calculation over and over, and no

Table 1 Orbital period of planets and Its average relative distance from the sun

	Distance to the sun P/Astronomical units	Orbital period around the sun R/Year	P^3	R^2
Mercury	0.3880	0.2409	0.0580	0.0580
Venus	0.7240	0.6160	0.3800	0.3795
Earth	1.0000	1.0000	1.0000	1.0000
Mars	1.5240	1.8809	3.5400	3.5380
Jupiter	5.2000	11.8631	140.6080	140.7300
Saturn	9.5100	29.4565	860.0850	867.6900

one else knew how hard it was except for Kepler. A few years later, after countless failures, Kepler finally found a simple and magical law: the square of the orbital period of the planet is proportional to the cube of its distance from the sun.

For this result, do you have a feeling of "Ah, if it were me, I might also be able to find this law!".

This is the third law of planetary motion, also known as the periodic law, an achievement that Kepler was most proud of.

The Copernican system won another great victory, because such a beautiful law can be obtained only based on the Copernican system, and there are no such beautiful laws in all geocentric systems.

Kepler's astronomy still has to face the traditional test. Can it be used as the basis for a more accurate star catalogue? Heliocentrism has to predict the motions of planets more accurately to be accepted by people with conviction. This star catalogue finally appeared in 1627, and its accuracy was surprisingly revealed four years later. From then on, strictly speaking, what astronomers call the Copernican system actually refers to the Copernicus-Kepler system.

A correct model may not be as accurate as a carefully refined wrong model in the beginning, and Copernicus' heliocentrism is just so compared with Ptolemy's geocentric system. However, the harmony, unity, nature and simplicity of the Copernican system convince many astronomers that the general direction of it is right. Kepler is such a great astronomer who made unremitting efforts for the general direction and achieved brilliant successes.

2 "Columbus" of Sky—Galileo

Galileo (Fig. 5) was born in Pisa, Italy in 1564 (seven years older than Kepler). In 1581, he entered University of Pisa to study medicine, but he was mainly interested in mathematics and physics. In 1589, when Galileo was 25, he was hired as a professor of mathematics at the University of Pisa and later transferred to the University of Padua to teach.

Galileo never believed in authority, including Aristotle's theories, which were praised as scripture. Aristotle was a great encyclopedic philosopher, scientist and educator in ancient Greece. His enormous knowledge, profound insights and outstanding contributions made him worthy of the title of "Great". However, no matter how great a person is, he cannot be perfect, and he cannot be right in everything that he says or does. Besides, countless examples in history tell us that once a character is respected as "Great", people usually hold all his words as scriptures, including his wrong words and right words that need to be corrected with the development of society. Moreover, Aristotle and his theories were used by the church and praised as the ideological authority of the Christian world.

It takes courage, knowledge and wisdom to question such an authoritative theory.

Aristotle once asserted that the speed of a falling object is proportional to its weight, that is, heavy things fall faster than light things. This is not from reasoning

Fig. 5 Galileo (The picture is from the Internet)

or experiment but by intuition, which is actually an illusion. It is said that Galileo once did experiments on the Leaning Tower of Pisa to prove it was wrong, but this story is unreliable. In fact, Galileo did not need to do any experiments, because he completely repudiated Aristotle's assertion in just a few words. He asked what would happen if a heavy object and a light object were tied together and fell from the sky. For example, if a big stone is tied to a small stone, according to Aristotle's assertion, there can be two answers. First, the falling speed of "combined stone" is between the falling speed of the big stone and the small stone, that is, it is more slowly than the falling speed of the big stone and faster than that of the small stone. Second, the falling speed of "combined stone" is larger than that of the big stone, because it is heavier than the big stone. The two answers are both "supposed" to be right, but they contradict each other, so Aristotle must be wrong. During as long as 1900 years, no one found the contradiction or thought of this simple reasoning. It was Galileo that found and pointed out Aristotle's mistake, and simply a few words made this assertion collapse, which aroused people's great admiration for his intelligence.

Aristotle believed that a force must be applied to the object before the object can move, and without force, the object has to stand still. This is also in line with people's intuition, because experience tells us that if we want to make a stationary object move, we have to apply external force to it. For example, we can push, pull, and lift it, or use something else to apply a force to it, and it is necessary to apply greater force to make it move faster, just like that a car can move faster when pushed by two persons than by one person alone. However, Galileo pointed out through experiments and reasoning that an object moving in the horizontal plane stops because of friction resistance. If there is no friction, once an object has a certain speed, it will keep moving at this speed. Movement does not require force to maintain. Galileo revealed Aristotle's mistake again, and gave a correct conclusion, which was later written into the law of inertia by Newton.

Galileo also studied the motion of projectiles. He found that the motion of a projectile can be divided into two parts. One is the uniform motion along the horizontal direction, and the other is the uniform accelerated motion in the vertical direction. He proved that the path of motion of projectile generated by horizontal uniform motion and vertical uniform accelerated motion is a semi-parabola.

Galileo's experiments and reasoning are explained in detail in physics textbooks of high schools. Einstein said that Galileo's contribution to science lies in destroying intuitive ideas and replacing them with new ones.

In the summer of 1609, Galileo (who had already firmly believed in the Copernican system) heard that a Dutchman had invented a telescope for people to enjoy. The telescope is mainly composed of two glasses. When the two glasses are fixed in front of the eyes from far to near, the distant scenery will get closer and bigger. Galileo was very excited by this news, so he decided to verify the truth of it very carefully and try to make such an instrument for himself. Galileo was not the inventor of the telescope, but he was the first to point the telescope to the vast sky. Since then, astronomy has obtained powerful observation tools, and human vision can go deep into the universe. Later, Galileo made one telescope after another, and every telescope was better than the last one. By the end of 1609, the magnification of his telescope had reached 20 times. Importantly, the telescope, in Galileo's hands, revealed countless evidence supporting the Copernican system.

At the end of 1609, Galileo observed the moon with his newly created telescope and found that the surface of the moon was uneven, with mountains and plains, as well as countless craters, and then he drew the first map of the moon personally. When he was observing stars, he found that they were much brighter than those observed with the naked eye, and there were many stars that could not be seen before. The Milky Way, originally a fog of light, has become a large area of countless stars. The telescope increased the number of visible stars in the sky, but did not increase their apparent size, and they were still small spots one by one. On January 7th, 1610, Galileo began to observe Jupiter with a telescope, and discovered that Jupiter was surrounded by four satellites.

At the beginning of 1610, Galileo published a book *The Starry Messenger* and published the above important discoveries, which caused a sensation.

In August 1610, Galileo observed Venus and discovered that Venus had the same phenomena of waxing and waning like the moon. At the end of 1610, he observed the sun again and found sunspots on it, and he calculated that the rotation period of the sun was 25 days according to the law of movement of sunspots. He wrote these observation results into the book *Letters on Sunspots* published in 1613.

Galileo's discoveries did not provide direct evidence to prove that the earth revolves around the sun, but all of them are strong support for the Copernican system. The uneven surface of the moon and the sunspots proved that Aristotle's assertion that there was a boundary between the heaven and the earth and celestial bodies were perfect was wrong, and it was right that there is no boundary between the heaven and the sky and the earth is also one of the celestial bodies. The telescope increases the number of visible stars in the sky, but does not increase their apparent size, which proves that stars are as far away from our earth as Copernicus predicted. Jupiter is found to be surrounded by four satellites, which proves that celestial bodies do not have to revolve around the earth, and the earth is not the only planet with satellites. The rotation of the sun undoubtedly reminds us that the earth can also rotate since the huge sun is rotating.

The discovery of Venus's waxing and waning is a heavy blow to the Ptolemaic system, which proves that Venus is indeed revolving around the sun as described in the Copernican system. As shown in Fig. 6, if the orbit of Venus is located between the sun and the earth as arranged by Ptolemy (Fig. 6a), we can always see only crescent Venus from the earth, but not full Venus. If Venus moves around the sun

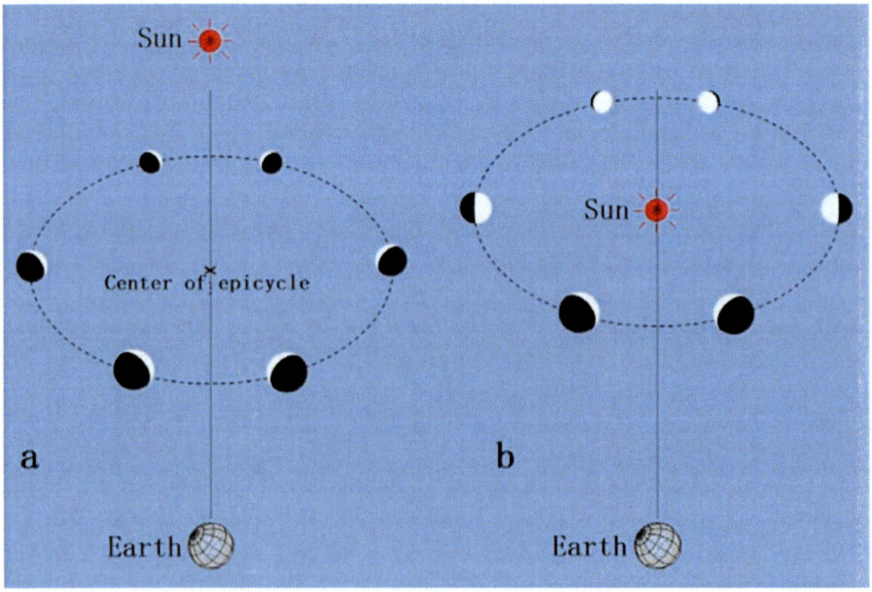

Fig. 6 Venus's waxing and waning (a) Venus's Revolution between the Sun and the Earth (b) Venus's Revolution around the Earth

as Copernicus arranged (Fig. 6b), we can definitely see the periodic waxing and waning of Venus. Copernicus could not see the waxing and waning of Venus, but the Copernican system predicted that this phenomenon must exist. Galileo's discoveries made people seem to hear Copernicus say, "I told you so!" This sentence is annoying, but no other sentence is more effective than it.

Galileo popularized astronomy with a telescope, and what he popularized was Copernican astronomy. His behavior aroused the vigilance of the church. In 1615 the Roman Inquisition issued a warning to those who supported the Copernican system. In early 1616, the Roman Inquisition arraigned 51-year-old Galileo and forbade him to teach or defend Heliocentrism orally or in writing. On March 5, 1616, the Roman Inquisition declared banning of Copernicus' *On the Revolution of Heavenly Spheres.*

Galileo, as a scientific fighter, did not give up his research and publicity. In 1632, he published the book *Dialogue Concerning the Two Chief World Systems.* He responded to almost all the criticisms and doubts about the Copernican system through the dialogue of three characters in the book.

Let's see how Galileo discussed the earth's diurnal rotation and its revolution around the sun.

While discussing the earth's diurnal rotation, Galileo said (through words of the characters in the book), "Whatever motion comes to be attributed to the earth must necessarily remain imperceptible to us and as if nonexistent, so long as we look only at terrestrial objects." "So the true method is to observe and consider whether bodies separated from the earth exhibit some appearance of motion which belongs equally to all." "Now there is one motion which is most general and supreme over all, and it is that by which the sun, moon, and all other planets and fixed stars—in a word, the whole universe, the earth alone excepted—appear to be moved as a unit from east to west in the space of twenty-four hours." "Each one incontrovertibly has its own motion from west to east, this being very gentle and moderate, and must then be made to rush the other way; that is, from east to west, with this very rapid diurnal motion. Whereas by making the earth itself move, the contrariety of motions is removed."

Galileo demonstrated in great detail that a series of arguments put forward by the opposition to prove the geostationary were untenable through experiments and geometric methods. For example, when the stones on the high tower fall, they always fall along a straight line parallel to the tower. The arrows shot vertically upward finally return to the original place. The shells fired with the same power have the same range no matter which direction they are fired in. None of these can prove that the earth is stationary, because if the earth rotates, the phenomena we observe will be the same, just like that there is no difference between throwing and jumping in a cabin that is sailing smoothly and throwing and jumping onshore. It is absurd to think that if the earth moves, the stones falling from the high tower will fall far to the west of the tower, the arrows shot vertically upward will fall far to the west of the archer, and the shells fired with the same power will fall very close if they are fired eastward and far if they are fired westward.

When discussing the annual motion of the earth, Galileo said, "I might very reasonably dispute whether there is in nature such a center, seeing that neither you nor anyone else has so far proved whether the universe is finite and has a shape, or whether it is infinite and unbounded. Still, conceding to you for the moment that it is finite and of bounded spherical shape, and therefore has its center, it remains to be seen how credible it is that the earth rather than some other body is to be found at that center." "The most palpable of these is that we find all the planets closer to the earth at one time and farther from it at another." "Mars, Jupiter, and Saturn are always quite close to the earth when they are in opposition to the sun, and very distant when they are in conjunction with it." "It is certain that Venus and Mercury must revolve around the sun, because of their never moving far away from it, and because of their being seen now beyond it and now on this side of it, as Venus's changes of shape conclusively prove." "And since it is true that all the planetary orbs (I mean Mercury, Venus, Mars, Jupiter, and Saturn) move around the sun as a center, it seems most reasonable for the state of rest to belong to the sun rather than to the earth."

Galileo said, "Annual motion, mixing with the individual motions of all the planets, produces a great many oddities which in the past have baffled all the greatest men in the world."

Galileo introduced in detail how the Copernican system explained planetary retrograde motion (explained in Chap. 7 of this book), and said, "The theory of this alone ought to be enough to gain assent for the rest of the doctrine from anyone who is neither stubborn nor unteachable."

The earth is revolving around the sun, "And such being the case, it necessarily follows that the diurnal motion, too, belongs to the earth. For if the sun stood still, and the earth did not revolve upon itself but merely had the annual movement around the sun, our year would consist of no more than one day and one night; that is, six months of day and six months of night, as was remarked once previously."

"See, then, how two simple noncontradictory motions assigned to the earth, performed in periods well suited to their sizes, and also conducted from west to east as in the case of all movable world bodies, supply adequate causes for all the visible phenomena." "I leave it to your judgment which has the more likelihood in it."

Galileo said, "There is no limit to my astonishment when I reflect that Aristarchus and Copernicus were able to make reason to conquer sense that, in defiance of the latter, the former became mistress of their belief."

The publication of *Dialogue Concerning the Two Chief World Systems* aroused great receptions, and the Copernican system was understood and accepted by more people. Galileo was the most powerful advocate of the Copernican system. Shortly after the publication of the book, the Roman Inquisition ordered a halt to the sale of the book and issued a subpoena for the trial of Galileo. At the beginning of 1633, Galileo, a sickly man, approaching the age of 70, came to Rome to stand trial after experiencing hardships with the help of his friends. Under great pressure, Galileo renounced Copernican doctrine in exchange for "forgiveness" from the Inquisition, and he was exempted from fire and was sentenced to life imprisonment. According to legend, Galileo muttered to himself after listening to the trial, "And it yet moves."

On January 8, 1642, Galileo passed away.

3 Newton Standing on the Shoulders of Giants

At the end of 1642, just after Galileo's death, Newton (Fig. 7) was born in Woolsthorpe in Lincolnshire, England. He was admitted to Trinity College of Cambridge University in 1661, but he did not obtain official status as a student. From 1665 to 1666, Cambridge University gave students a long holiday because of the local plague, which happened to be the period of the extraordinary peak of Newton's intelligence creation. When Cambridge University resumed classes in 1667, Newton returned to school. His teacher Barrow appreciated Newton's talent very much. In 1669, Barrow gave his professorship to Newton, who was under 27 years old.

Newton summed up a series of great discoveries in mechanics and astronomy at that time. Based on the work of scientists such as Galileo and Kepler, through his own practice and research, he sorted out a systematic theory and established Newton's mechanics system by means of mathematical analysis.

In 1687, Newton's *Mathematical Principles of Natural Philosophy* came out. In this great work, Newton expounded his three laws of motion and the law of universal gravitation. The formulas and proofs of Newton's laws of mechanics are explained in detail in physics textbooks of high schools and universities, so the following is only a brief mention.

First law: an object either remains at rest or continues to move at a constant velocity, unless acted upon by an external force.

Fig. 7 Newton (The picture is from the Internet)

Second law: the acceleration of an object as produced by a net force is directly proportional to the magnitude of the net force, and inversely proportional to the mass of the object.

Third law: all forces between two objects exist in equal magnitude, opposite direction and the same straight line.

Law of universal gravitation: every particle attracts every other particle in the universe using a force that is directly proportional to the product of their masses but also inversely proportional to the square of the distance between them.

Based on these laws, Newton explained why the moon revolves around the earth in a circular orbit. He said that all objects orbiting in any orbit attempt to leave their orbital center. If there is no opposing force to curb their attempts and restrain them in orbit, they will fly away at a constant speed along a straight line. This opposing force is exactly gravity.

Newton made a thought experiment (Fig. 8). If a shell is launched from the top of the mountain and the launching direction is parallel to the ground plane, the shell will fly along a curve for 2 miles before landing. If the power of gunpowder is increased to double the firing speed or increase the firing speed to ten times, the flight distance of the shell will double or increase to ten times (assuming no air resistance). By increasing the power of gunpowder to increase the launching speed, the shell's flying distance can be increased at will, and the curvature of its trajectory can be reduced, so that it can fly far away, even around the earth without falling. If the power of gunpowder is increased again, the shell will never return to the earth, but fly straight into outer space and make an infinite motion.

By the same token, the reason why the moon moves around the earth in a circular orbit instead of flying away at a constant speed in a straight line is that the moon is constantly pulled toward the earth by gravity, deviating from the straight path followed by its inertial force and moving along its current orbit. If there is no such force, the moon will not be able to stay in its orbit. If this force is too small, it will not be enough to make the moon deviate from the straight path, and the moon will fly away from the earth. If the force is too large, it will cause too big deflection and pull the moon from its orbit down to the earth. Therefore, this force must be in an appropriate quantity, and the mathematician must find the force that makes an object move accurately along a given orbit at a given speed.

Similarly, the reason why the planets revolve around the sun is also "the combination of gravity and propulsion". Newton extended the laws of motion applied on the earth to the sky, proving that the laws of nature are effective for the whole universe, and the world has the unity of matter.

Kepler's three laws of planetary motion have not been widely recognized until Newton used them as the basis of his cosmic system theory. Newton proved a series of propositions related to Kepler's law. It also can be said that Newton deduced these propositions. These abstract results from Newton's mechanical perspective can be applied to the solar system, so as to obtain the true laws of the motions of celestial bodies in the solar system. Until this time, people no longer doubted Kepler's three laws of planetary motion. Until this time, people understood the seemingly strange orbits of comets, and comets were incorporated into a unified system.

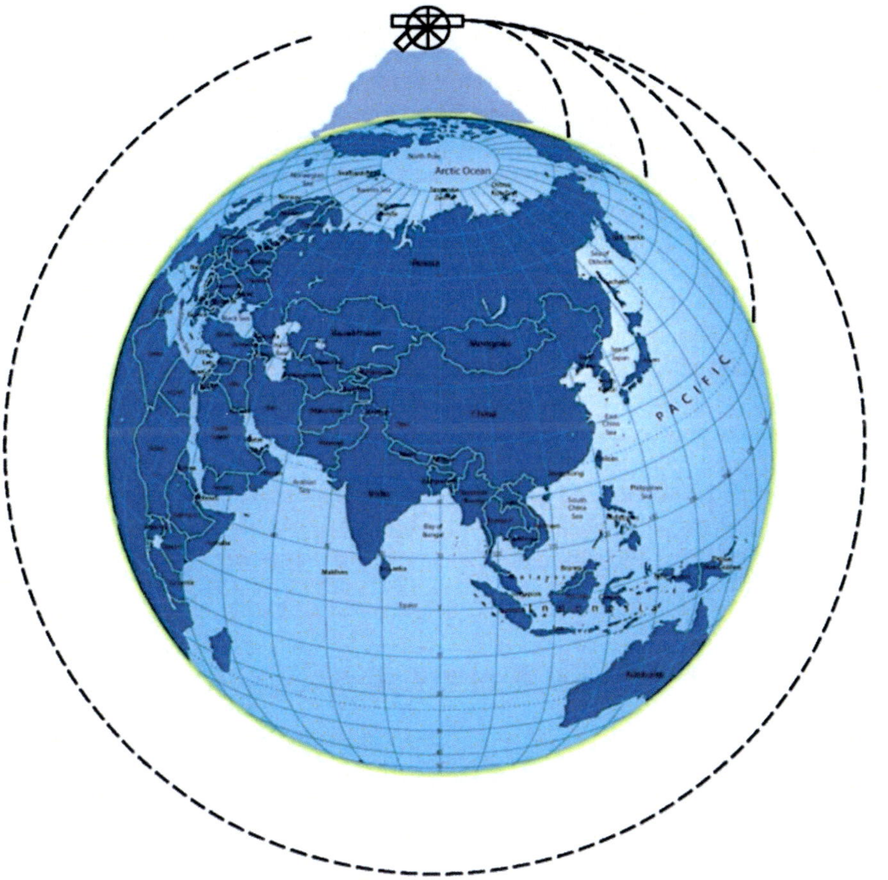

Fig. 8 Newton's thought experiment

Kepler's laws are the products of experience, which is inferred from Tycho's careful observation, or even "scraped together". Newton's laws are theoretical and they are simple mathematical concepts, based on which we can finally deduce all the data observed by Tycho.

Newton found that when a projectile is moved by a force that points to the focus and is inversely proportional to the square of the distance, its orbit is not only limited to Kepler's ellipse, but also can be any kind of conic curve. He deduced various properties of motions on these curves, and pointed out that the necessary conditions for determining the curve shape of the orbit (circle, ellipse, parabola or hyperbola) are limited to the initial conditions of objects. For example, the orbit of the earth is elliptical while moving at a speed of 29.8 km/s in its current position. If the speed is increased to 42 km/s, its orbit will become an open parabola, and then it will break away from the control of the sun's gravity and go beyond the solar system. Newton extended Kepler's first law.

Just as Pierre-Simon Laplace said, "Without it, the ellipticity of the planetary orbits, the laws which the planets and comets follow in their revolution around the Sun, the precession of the equinoxes, the nutation of the terrestrial axis, the motions of the lunar axis, and lastly, the ebbing and flowing of the sea, would only be insulated and unconnected phenomena. It is really a circumstance deserving our admiration, the manner in which all these phenomena, at first sight so unconnected, flow from one law."

With Newton's discovery of universal gravitation, people finally understand why people on "the other side" of the earth did not "fall", and why objects on the earth's surface could not be thrown out by high-speed rotation of the earth. Copernicus' theories became more convincing when the research on the role of force in motions of celestial bodies became a part of astronomy, because, in this theory, planets revolve around a huge sun instead of the much smaller earth. Copernicus' astronomy became credible in both physics and cosmology for the first time. The law of the relationship between the earth and other objects in the universe was formulated once again, and the last substantial opposition of planetary earth disappeared.

4 Evidence of the Earth's Revolution

Astronomers still hoped to find direct and irrefutable evidence though the theories had been very convincing, and they firmly believed that such evidence must exist. With the development of science and technology, some important evidence was discovered one after another. In the following description, there are some professional terms, which will not be introduced in detail in this book. Readers need to read professional books if they want to study deeply.

1. Discovery of Aberration

According to the Copernican system, when the earth moves in an orbit around the sun, it moves from one end of the orbit to the other within 6 months. The distance is so great that the annual parallax of stars will be produced definitely. To prove the Copernican system, there is nothing more convincing than one or two examples of stellar parallax.

James Bradley, a British astronomer and director of Greenwich Observatory in the early eighteenth century, was a scholar who devoted himself to finding the annual parallax of stars. In 1725, he used a vertically installed telescope (zenith instrument) to observe a Gamma Draconis passing through the zenith of Greenwich. Every day (except for a time when it was too close to the sun), this star passed through the mirror of the telescope. He found that the star would move slightly when passing through the north–south marking line in the center of the telescope, from north to south and then from south to north. The moving period is one year, and the moving distance in the north–south direction is only 40 arc seconds, which is extremely small. It was undoubtedly the result of the earth's movement around the sun, but to Bradley's surprise, the direction of this displacement was not consistent with the expected parallax displacement. Obviously, it was not the annual parallax of stars, but another physical effect, and Bradley thought hard about it.

On one occasion, when he sailed on the River Thames, he found that the flag on the top of the mast did not simply sway with the wind, but changed its direction according to the relative motion of the ship and the wind. This situation is the same as that of people walking in the rain with an umbrella. If the rain falls vertically, the raindrops will fall on a pedestrian if he holds an umbrella vertically above their heads. If the umbrella is tilted forward slightly, he will not get caught in the rain. Moreover, the faster a person walks, the more the umbrella must tilt forward. The umbrella should tilt slightly eastward if he goes east and southward if he goes south, and it is the same when he goes west or north. Bradley suddenly realized that light falls on the earth from a star in a certain direction and at a certain speed, while the earth is orbiting the sun at another speed. Like an umbrella, the telescope must tilt slightly in the moving direction of the earth, so that the light of a star can fall straight on the lens. Bradley called this inclination angle "aberration".

The aberration refers to the difference between the apparent directions of celestial bodies observed by the moving observer and the true directions of celestial bodies observed by the stationary observer at the same moment.

Bradley hoped to find the stellar parallax, but what he found was the aberration. The accidental discovery of the aberration provided another proof that the earth was moving around the sun.

2. Accurate Prediction of the Return of Halley's Comet

In the *Mathematical Principles of Natural Philosophy*, Newton once made it clear that the comet's orbit extends very long and was around the sun, and its orbit is similar to a parabola. Comets leave the sun in an elliptical orbit and are destined to be pulled back into the solar system one day.

Halley, who is 14 years younger than Newton, is a famous British astronomer and a good friend of Newton. In the study of comet's orbit, he found that there were many similarities in comet's orbit observed in 1531, 1607 and 1682. In 1695, Halley told Newton that he thought they must be the same comet appearing in different periods. He predicted that the comet would return "at the end of 1758 or the beginning of the 1759".

In June 1757, with the help of two assistants, Clairaut revised Halley's prediction after careful and arduous calculation: the comet would orbit around the sun in mid-April, 1759, lasting one month. On January 21, 1759, Messier, a comet hunter in Paris, observed the returning comet. On March 13, the comet entered the orbit of the sun, and its orbital characteristics were very similar to those of the comets in 1531, 1607 and 1682. They are the same comet.

Halley passed away in 1742. People named this comet Halley's comet to commemorate his achievements.

Halley was proved to be right, and Newton's theories won the most extensive victory among the public, which was also the victory of the Copernican system.

3. Discovery of the Stellar Parallax

The search for stellar parallax was still going on. With the constant emergence of a new generation of instrument manufacturers, the accuracy of astronomical instruments has been continuously improved.

In 1835, German astronomer Wilhelm Struve chose α Lyra (Vega) as his research object. In 1837, he announced that from 17 observations, it can be inferred that this star has the parallax of 1/8 arc second, which is quite close to the modern value.

In 1834, German astronomer Friedrich Wilhelm Bessel began to observe 61 Cygni, but soon lost interest in it because of the arrival of Halley's comet. It was not until 1837 that he resumed his observation of 61 Cygni. He spent more than one year on intensive observation of this star. The number of repeated observations per night was as high as 16 times under normal circumstances, and he observed even more frequently when the visibility was particularly good. By the end of 1838, he announced that the parallax of the star was about 1/3 arc second. His results were convincing because his multiple observation patterns were consistent with theoretical expectations.

Just a few weeks later, Thomas Henderson, a British Royal astronomer who once worked at the Royal Observatory, Cape of Good Hope, announced that the parallax of a star in the southern celestial sphere exceeded 1 arc second, which was Alpha Centaurus. But this value is not very accurate. According to later measurement, the annual parallax of Alpha Centaurus is 0.76 arc second.

The discovery of stellar annual parallax is an outstanding achievement in the history of astronomy, and it is the most powerful proof of the earth's movement. The annual parallax of the star is so small, which indicates that stars are really very far away from us. Alpha Centaurus is the nearest star to us, and its distance from the sun is about 270,000 times that between the sun and the earth, which is really unimaginable, and this was predicted by Copernicus.

4. Calculated Neptune

In 1781, Wilhelm Herschel, an organist and amateur astronomer in England, discovered a "strange" star when he observed the stars with a self-made reflective telescope. This star presented a small circular surface in the telescope, which is different from fixed stars. The telescope could not magnify stars, but could enhance their brightness and turn them into bright spots. Subsequent follow-up observation found that the "strange" star moved slowly in the stellar background. According to the observation data, astronomers analyzed and concluded that this was a planet moving around the sun, which was farther away from the sun than Saturn. This was the seventh planet in the solar system, and it was named Uranus according to mythological tradition.

The magnitude of Uranus is 5.7. Although it is very dark, it is still in the visible range of human eyes. Unfortunately, no one has noticed it for thousands of years, and it was not captured by a telescope until 1781. Conceivably, it is too difficult to find a very dark and very slow planet in the vast starry sky. It can also be imagined how patient and persistent astronomers are to observe the stars.

Soon, Uranus's orbital characteristics were calculated and its position table was worked out. However, Uranus deviated from its expected orbit many times, which puzzled astronomers, so a variety of explanations have sprung up, two of which were worth considering: perhaps the law of gravity would deviate from the inverse square form at large distances; or Uranus was attracted by an undiscovered planet in its periphery. In the mid-eighteenth century, people tried again and again to modify the law of gravity, but no matter how hard they tried, the law was always as firm as a rock. There hence was a consensus that Uranus's abnormal behavior was the result of the action of an undiscovered planet.

How to find this unknown planet in the vast ocean of stars?

Search the sky with a telescope like Herschel? But the unknown planet is more distant, dimmer and slower than Uranus, and search the planet one by one is no different from looking for a needle in a haystack. Astronomers were more inclined to use Newton's mechanics theory to calculate the position of this planet within the framework of the Copernican system according to the clues provided by Uranus.

Adams, a graduate of University of Cambridge, and Le Verrier of Paris Observatory, France, almost simultaneously and independently carried out complicated and difficult calculations for the position of this planet. In October 1843, Adams got the approximated solution of this star's position, but he did not get a more accurate result until September 1845 due to his busy teaching schedule. He calculated that the planet should be at the longitude of $322^\circ\ 34'$ centered on the sun on October 1. In France, in June 1846, Le Verrier concluded in a paper that the longitude of the planet must be around 325° when viewed from the sun.

Adams' work was delayed by the royal astronomer Airy of Greenwich and James Charles, a professor of astronomy at the University of Cambridge, not being observed and confirmed in time, while Le Verrier persuaded the astronomers at Berlin Observatory to search the star soon after he reached his conclusion. On September 18th, 1846, Le Verrier wrote to astronomer Galle of Berlin Observatory with a detailed star chart, asking for help in searching for the star. He wrote in the letter, "Please point your telescope at 326° of the zodiac in Aquarius, and you will find a new planet with an obvious round surface in the area about 1° away from this point. Its magnitude is about 9…".

Le Verrier's letter arrived in Galle's hands on September 23, and Galle aimed the telescope at the small area designated by Le Verrier that night. It was found that there was indeed a star that was not on the star chart. This was exactly the planet that Le Verrier wanted to find, and it was located $52'$ away from the point predicted by Le Verrier. The planet moved the next night, and Galle immediately wrote back to Le Verrier in Paris on September 25th, "Sir, the new planet you pointed out to us is real."

The discovery of a new planet caused a sensation in the world, which is a great miracle in the history of science, and an excellent example of the great foresight of scientific theories, and an unparalleled great victory of Copernican-Newton theory.

Astronomers named the eighth planet in the solar system Neptune. The history of science also recorded the achievement of Adams and Le Verrier fairly.

Engels said, "For three hundred years the Copernican solar system was a hypothesis with a hundred, a thousand or ten thousand chances to one in its favour, but still always a hypothesis. But when Le Verrier, by means of the data provided by this system, not only deduced the necessity of the existence of an unknown planet, but also calculated the position in the heavens which this planet must necessarily occupy and when Galle really found this planet, the Copernican system was proved."

Epilogue

Celestial bodies are either too close to us or too far away from us. Human beings are observing the earth with their tiny bodies close to the huge earth, and standing on the moving earth to observe other distant celestial bodies that are also moving, in which the hardship is conceivable.

Who has seen the earth revolve around the sun? Now, we have already completed this proof together with astronomers in history when we raise this question again. We are capable of knowing that the earth is spherical without leaving the earth; we can be sure that the earth is revolving around the sun without standing perpendicular to the earth's orbit plane and hundreds of millions kilometers away from the center of the sun and observe for a whole year. This is human wisdom!

With such wisdom, human beings not only know the solar system, but also have gone far beyond it, extending the tentacles of exploration to the depths of the vast universe.

Many things in nature are difficult to see directly with naked eyes. For example, who has seen electrons revolving around the atomic nucleus? Who has directly seen the double helix structure of DNA? No one. Maybe one day in the future, scientists can directly observe the structure of atoms and DNA through some equipment, but they do not need to wait for that day to come, because they are able to demonstrate and describe them through experiments and reasoning. There are so many such examples, and this is human wisdom.

Infinite human wisdom!

Y. Ge, *From Geocentric to Heliocentric: How Discoveries Are Made*,
https://doi.org/10.1007/978-981-16-7779-3

Postscript

When I was a child, I was quite curious. How far are the stars in the sky from us? The sun sets in the west every day and rises in the east the next day. How does it get through the ground? How big is the ground? Will I reach the end of it if I keep going straight ahead? Is there anything blocking the edge of the ground? A fence? Then what is outside the fence? The more I thought, the more scared I was.

As I read and grew, I understood that the ground under our feet is a sphere, that the earth is revolving around the sun, and why people on "the other side of the earth" do not fall. Later, I also learned to identify the stars in the night sky, understood what the ecliptic is and why the planets would move backward.

But new questions arose, and I was curious. The stars are invisible when there is a sun, so how do astronomers know the sun's motion in the background of stars? And how can they know its accurate motion path? Then I also wondered. In an age when people's scope of activities was rather small and there were only simple technologies and tools, how did they know that the earth was a sphere? I know that the heliocentric theory was put forward by a man named Copernicus more than 400 years ago, but even the telescope had not been invented at that time, let alone the spaceship or the plane. How did he realize that the earth was revolving around the sun with the naked eye?

When I was in college, I spent half a year in the library searching for information and exploring the answers that I wanted to know, which is hard but fun. I basically found the answers I needed and satisfied my curiosity. More importantly, I saw the unremitting exploration spirit of those great scientists and philosophers in ancient times and their outstanding wisdom, which benefited me a lot.

The history of human exploration of the truth of the solar system is vivid, interesting and rich in connotation, which can give people a lot of enlightenment. I think if I tell it to junior high school students, they will be very interested, won't they?

Although I had a thought, after all, this is not my major, let alone my career. Therefore, I did not focus on it in the next 30 years.

Ten years ago, when I was about to retire, I could do something with my interest in my spare time. I thought of it again and decided to write a book to write down the history of mankind's early exploration of the structure of the universe. I hope this book can be useful to junior high school students and others interested in it.

However, I met many challenges in the process of writing. There is no need to say more about the hardship. In short, after suspending and resuming over and over again, the book is finally completed, and it is up to the readers to judge how it is written.

I am very grateful to Professor Zhang Pengjie of Shanghai Jiao Tong University for his preface to this book, and to Professor Chen Xuelei of National Astronomical Observatory, Professor Li Weigu of Peking University, Professor Zhang Ping of Tel Aviv University of Israel, Professor Qian Hong of Washington University in Seattle, USA, and Professor E Weinan, academician of Chinese Academy of Sciences for their recommendations. I am extremely touched by their enthusiastic support. I am very grateful to Editor Hou Junlin and other editors of Science Press for their requirements and suggestions on this book, and for their efforts in publishing this book. I am very grateful to Professor Bai Shunong, Professor Huang Yanyi and Professor Gao Yiqin of Peking University for their support and help to this book. I would also like to thank Professor Ge Hao of Peking University for carefully reading my manuscripts time and time again and putting forward many valuable comments. At the same time, I would like to thank Wang Zhuqin, Xu Shidong, Tan Qinghe and other friends who have read my manuscripts and put forward many valuable comments.

Ge Yunbao (葛云保)
March 2015 in Beijing

Bibliography

1. Shi YL (1996) *History of Science and Technology in Ancient China (Astronomy)*. Liaoning Education Press, Shenyang
2. Zhang Yuzhe (ed.) (1980) *Encyclopedia Of China (Astronomy)*. Encyclopedia of China Publishing House, Beijing
3. Jin QP (1997) *History of Ancient Chinese Science and Technology*. The Commercial Press, Beijing
4. The book's writing team (1981) *History On Astronomy*. Shanghai Scientific and Technical Publishers, Shanghai
5. Chen ZG (1980) *History Of Chinese Astronomy (Vol.1)*. Shanghai People's Publishing House, Shanghai
6. Chen ZG (1982) *History Of Chinese Astronomy (Vol.2)*. Shanghai: Shanghai People's Publishing House, Shanghai
7. Chen ZG (1984) *History Of Chinese Astronomy (Vol.3)*. Shanghai: Shanghai People's Publishing House, Shanghai
8. Cui ZH, Chen D (1993) *History Of World Astronomy*. Jilin Education Press, Changchun
9. Xuan HC (1992) *History Of Astronomy*. Higher Education Press, Beijing
10. Chen ZW (1980) *From Copernicus To Newton*. Science and Technology of China Press, Beijing
11. Deng KH (2009) *Tracing To The Origin Of Greek Mathematical Astronomy-- A Comparative Study Of Ptolemy's Almagest*. Shandong Education Press, Jihan
12. Deng KH (2011) *The History Of Chinese Astronomy From A Comparative Perspective*. Shanghai People's Publishing House, Shanghai
13. Jiang XY (2005) *Zodiac and Twenty-Eight Mansions—Astrology In History*. Liaoning Education Press, Shenyang
14. Jiang XY (1985) *Determination And Calculation of Positions Of The Sun In Ancient China*. Annals of Shanghai Observatory Academica Sinica, No.7: P91- P96
15. Jiang XY (2008) *It Is Time To Talk About Ptolemy* (1–5). Sina Blog (2008.4.9–2008.7.10) http://blog.sina.com.cn/jiangxiaoyuan.
16. Chen JJ (1978) *The Origin Of Calendar And Pre-Qin Quarter Remainder Calendar*. Collected Works on the History of Science and Technology. No.1 (Astronomical History): P5-P21.
17. Chen JJ (1978) *A New Probe Into The Development History Of Theory Of Sphere-Heavens*. Collected Works on the History of Science and Technology. No.1 (Astronomical History): P59-P74.
18. Barnes J (ed.) (1984) *The Complete Works of Aristotle*. Princeton University Press, New Jersey
19. Sarton G (2012) *Hellenistic Science And Culture In The Last Three Centuries B.C.* Translated by Lu XD, etc. Elephant Press, Zhengzhou
20. Katz VJ (2004) *History Of Mathematics: An Introduction. Second Edition*. Translated by Li WL, etc. Higher Education Press, Beijing

Y. Ge, *From Geocentric to Heliocentric: How Discoveries Are Made*,
https://doi.org/10.1007/978-981-16-7779-3

21. Kuhn TS (1957) *The Copernican Revolution: Planetary Astronomy In The Development Of Western Thought*. Harvard University Press, Cambridge
22. Laplace PS (2007) *The System of the World.* Translated by Pond J. Kessinger Publishing, White Fish, Montana
23. Newton I (2001) *Newton's Philosophy Of Nature Selections From His Writings.* Translated by Wang FS, etc. Shanghai Century Publishing Group Shanghai Translation Publishing House, Shanghai
24. Newton I (2001). *Mathematical Principles Of Natural Philosophy.* Translated by Wang KD. Shanxi People's Publishing House, Wuhan Press, Xi'an
25. Copernicus N (1992). *On the Revolutions Of the Heavenly Spheres.* Translated by Rosen E. Johns Hopkins University Press, Baltimore
26. Galilei Galileo (1967) *Dialogue Concerning The Two Chief World Systems: Ptolemaic and Copernican.* Translated by Stillman Drake, Berkeley and Los Angeles
27. Hoskin M (eds.) (1996) *The Cambridge Illustrated History of Astronomy* (Cambridge Illustrated Histories) Cambridge University Press, Cambridge
28. Sagan K (2002) *Cosmos.* Random House, New York
29. Luo GZ (2002) *Romance of Three Kingdoms.* Translated by Brewitt-Taylor CH. Tuttle Publishing, Clarendon
30. Aristotle (1924) *Metaphysics.* Translated by Ross WD. Oxford University Press, Oxford
31. Marx K and Engels F (1975) *Marx/Engels Collected Works Volume* 4. International Publishers, New York